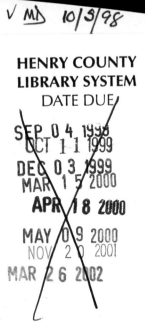
PURCHASE MADE POSSIBLE BY

GOVERNOR ZELL MILLER'S
READING INITIATIVE

1998

RODENTS
OF THE
WORLD

RODENTS
OF THE
WORLD

David Alderton

Photographs by Bruce Tanner

Facts On File®

A BLANDFORD BOOK
First published in the UK 1996 by Blandford
A Cassell Imprint

Cassell Plc
Wellington House
125 Strand
London WC2R 0BB

First published in the United States 1996 by
Facts On File, Inc.
11 Penn Plaza
New York, NY 10001

Library of Congress catalog card number 96-15285

Facts On File books are available at special discounts
when purchased in bulk quantities for businesses,
associations, institutions, or sales promotions.
Please call our Special Sales Department at
212/967-8800 or 800/322-8755.

ISBN 0-8160-3229-7

Typeset by York House Typographic Ltd., London

Printed and bound in Spain by Bookprint

Previous pages: The light colour of the fat sand rat (*Psammomys obesus*) helps to conceal its
presence in the desert areas of North Africa where it lives.

Contents

Acknowledgements

The photographs were taken by Bruce Tanner, apart from those supplied by the author (pages 31, 43, 170 top), Graham Thurlow (pages 97, 109, 111 top, 113 top) and the US Fish & Wildlife Service (pages 106 M.D. Spanel; 110, 114 Ron Singer; 115 Jim Leupold; 119 Lloyd Poissemot; 124 George Harrison; 132 R. Town; 135 C. McCaffery).

Thanks are due to the many people who assisted with the preparation of this book, with a particular debt being due to the following:

Michael Boddington, Sussex, England.
Martin Bourne, Lancashire, England.
Jim and Sandra Collins, Lancashire, England.
Custom Aquaria, Buckinghamshire, England.
Josephine Doolan, London, England.
Jenny Gilchrist, London, England.
Sheila Griffiths, Surrey, England.
Chris Henwood, Sussex, England.
Gavin Hyde, Buckinghamshire, England.
London Borough of Lambeth Public Health and Pest Control Services, London, England.
Marwell Zoological Park, Hampshire, England.
Graham Thurlow, Yorkshire, England.
US Fish and Wildlife Service, Washington D.C., USA.
Photographic assistance: Denise Palmer.

Special thanks are due to Jim Collins, for his helpful comments on the manuscript, as well as Rita Hemsley and Sarah Sell for their contributions to the production of this book.

Preface

Both numerically and economically, rodents are without doubt the most significant group of mammals on this planet. They have conquered the entire world, in terms of their distribution, with the house mouse (*Mus musculus*) being present on all continents, including Antarctica, where it was introduced with supplies for the bases and is now thriving.

Their highly adaptable nature has been responsible for their success. Rodents live both on and under the ground, as well as conquering the skies and even adopting a predominantly aquatic existence.

They range from the blind mole rats (Spalacinae) which live entirely in subterranean tunnels, excavated by means of their greatly enlarged incisor teeth, to the various flying squirrels, which are able to glide easily from tree to tree. Then there are rodents found in water, such as the beavers (*Castor* species), whose remarkable engineering capabilities are well known. Their dams, built over successive generations, may measure more than 700 m (2,300 ft) across and can be strong enough to support the weight of a horse and rider.

Yet in spite of this remarkable adaptability, relatively few rodents are popular with people. Their harmful activities, ranging from the destruction of food stores to their role as the carriers of plague and other deadly diseases, have meant that this group of mammals is unlikely to engender a sympathetic response in conservation-campaigning terms.

Nevertheless there are rodents which desperately need human assistance in order to survive. Thankfully, captive breeding can help in many cases, provided that it is not left too late. Much still remains to be learnt about the life styles of many species, however, and there are doubtless new species of rodent still to be discovered in remoter parts of the world. This is a group of mammals full of fascination, whose adaptability and resourcefulness rivals our own.

David Alderton,
Brighton, England.

Chapter 1
Rodents and People

There are approximately 1,500 different rodents known to science, representing about 40 per cent of all the world's mammalian species. They range in size from the tiny harvest mouse (*Micromys minutus*), which averages about 6 g (¼ oz) in weight when adult, to the capybara (*Hydrochaeris hydrochaeris*), which can weigh nearly 80 kg (176 lb). Their success is primarily due to their highly adaptable nature. They are found in a wide variety of habitats, both on and under the ground, in trees, and in lakes and rivers, and have even mastered the air.

Their feeding preferences are equally diverse, ranging from plants and seeds to insects and fish, but, in spite of such diversity, rodents have retained a fairly distinctive appearance and their relatively small size is also advantageous for animals which tend to be heavily predated by many other mammals and birds. It helps them to either remain hidden or escape out of reach, into a hollow tree, for example, when danger threatens.

Many rodents show a natural tendency to increase their numbers rapidly under favourable conditions. Not surprisingly, therefore, some species will often be found living in association with people, although

The capybara (*Hydrochaeris hydrochaeris*) is the biggest of all living rodents. It is about the same size as a largish dog. It occurs in both Central and South America.

they are more likely to be betrayed by their droppings than actually sighted, unless they are present at very high densities. The nocturnal nature of many rodents, coupled with their size and coloration, helps them to escape detection.

Three particular species, the house mouse (*Mus musculus*), the black rat (*Rattus rattus*) and the brown rat (*R. norvegicus*), have expanded far outside their natural ranges, thanks to human activities, and have become major pests in the process. The house mouse originated in the dry steppelands of the former USSR. It had already managed to reach Britain by the Iron Age, in about 1200 BC, and has been of particular interest to scientists, who have been able to follow its evolution in some areas.

Populations of house mice on the Faeroe Islands, which are located between Iceland and Norway, share characteristics with the mice that occur on the Shetland Islands and the now extinct subspecies (*M. m. muralis*), which formerly lived on St Kilda. All these mice are thought to be descended not from mainland British stock but from others brought to these islands by Viking invaders from Scandinavia about a millennium ago.

Even more interesting is the way in which these isolated populations have subsequently diverged while still showing traces of a common ancestry. Their relationship can be confirmed by the shape of the

Alert and watchful, like this green acouchi (*Myoprocta acouchy*), rodents are often shy by nature and retreat to cover at the slightest hint of danger.

Plant matter, such as seeds, roots and tubers, as well as leaves and stems, are eaten by many rodents, such as this brush-tailed porcupine (*Atherurus* species).

mesopterygoid fossa, situated at the back of the skull, which usually narrows to a point.

It was in the early 1900s that the significant increase in size of the mice resident in the Faeroes first attracted scientific interest. Those established on the island of Nølsey were noted to be considerably larger than those found anywhere else. It has been suggested that greater size evolved because the mice could then scramble over the rocky cliffs more easily, among the colonies of nesting sea birds. Although it had been generally accepted that a time interval of approximately 5,000 years was required for a new subspecies to evolve, that following the introduction of these particular mice was clearly much shorter.

Wherever boats have carried cargoes from Europe, house mice have become established soon afterwards. Other signs of adaptation can be seen in mice from other islands around the world. On the island of South Georgia, near Antarctica, house mice survive in one of the most inhospitable areas of the planet. Temperatures are likely to be below freezing for at least 5 months of the year and blizzards can sweep over the island at any time.

The mice there inhabit the thick bases of tussock grass and have up to three times as much brown fat stored in their bodies as other colonies living nearer the equator. This type of adipose tissue, which is a particularly concentrated store of energy and vital for thermal insulation,

The destructive capabilities of rats (*Rattus* species) is such that they can gnaw through almost anything in their way, including power cables, which can result in serious fires.

is present in the subcutaneous layer of the skin. The breeding season in South Georgia is limited to the warmer months of the year, between October and April, whereas normally house mice can breed at any time.

Like those found in the far north, the size of the mice present on South Georgia has also increased, in accordance with Bergmann's Rule. This states that the body size of a particular taxon will increase in proportion to a fall in the mean temperature of their environment. This adaptation is of particular significance in the case of rodents, whose small volume, relative to their surface area, leaves them vulnerable to hypothermia.

House mice are equally adaptable in terms of altitude, living up to 4,570 m (15,000 ft) in the Andes, where it can also be very cold. They have managed to penetrate to the fringe of deserts, where there is little rainfall, and live equally successfully in some of the wettest parts of the world.

In certain areas, predatory species have benefited from the introduction of these mice. On the Galapagos Islands, for example, local races of both the short-eared owl (*Asio flammeus*) and barn owl (*Tyto alba*) have increased in numbers as a result of the more ready availability of rodent prey.

RODENTS AND THE SPREAD OF DISEASE

While mice can spread a wide variety of diseases, of which some, such as plague, pseudotuberculosis and yersiniosis, are potentially transmissible to people, rats have an even worse reputation in this respect. Both the

black rat (*Rattus rattus*) and brown rat (*R. norvegicus*) are thought to
have originated in South-East Asia and then spread, with human assistance, to all continents apart from Antarctica. They are also present on
many islands, where they have caused serious harm to indigenous species.

It used to be assumed that black rats first arrived in Europe as a result
of the Crusades in the Middle Ages, having been brought back by the
returning knights and their retinues. However, recent archaeological
evidence has shown that these rats were already present in Britain during Roman times.

The first major outbreak of bubonic plague also occurred during this
era. This was during the sixth century and was called the Plague of
Justinian, after the Roman emperor of that time. Prior to this, sporadic
localized epidemics of plague were commonplace. In about 1200 BC,
the Egyptian pharoah Rameses V probably fell victim to this rat-borne
infection, as evidenced by an ulcer on his body that survived the mum-
mification process.

It is likely that the decline of the Roman Empire was at least partly the
result of plague, which occurred at quite regular intervals. In AD 68,
there were 10,000 deaths each day as bubonic plague swept through
Rome. Just under a century later, almost the entire Roman army, which
had brought the infection back from Mesopotamia, and half the population of the capital, including two emperors, were dead. This outbreak
lasted for 16 years. It is likely that, during the Roman Empire, more
than 100 million people died of bubonic plague.

For reasons which are not entirely clear, outbreaks of plague then
became less common, but the disease did not disappear. The roots of
the second pandemic, which was ultimately to give rise to the Black
Death, have been traced back to the Gobi Desert. In this region the population of rats began to multiply very rapidly and, from here, started to
spread bubonic plague across Asia.

The disease was first recorded in Europe in 1347, after desperate
Turkish soldiers, who were attacking a port, now called Feodosiya, in the
Crimea, catapulted bodies of their dead companions into the city.
Fleeing on ships accompanied by black rats, the Genoese merchants
unwittingly carried the plague to Italy, where it ultimately killed half the
population.

The Black Death, as it became known much later in 1823, wiped out
more than 25 million people in Europe and an estimated 75 million in
total throughout Europe, Asia and Africa. It reached London in 1348,
killing nine out of every ten inhabitants. The effects were particularly
severe in towns, where people lived together in overcrowded and insan-
itary conditions. Swellings in the groin or armpits were an early symp-
tom. These rapidly spread over the body, turning blackish, which is why
the term 'Black Death' was coined to described this epidemic.

Panic swept the streets as people tried to avoid the deadly plague. A
wide range of often bizarre and unpleasant remedies were recom-
mended to guard against the disease. Billy goats were kept in people's
homes in the hope that their strong odour would purify the air. At this

stage no one appreciated that the infection was spread directly by the fleas of rats. A legacy of that grim period in history, however, is eau-de-Cologne. This scent was first formulated to protect against the plague.

Perhaps strangely among the wide range of causes put forward to explain plague, both dogs and cats were considered to be likely suspects and were killed as a result. This could only have worsened the situation, by destroying the rats' only natural enemies in the urban environment.

Children still recite the old nursery rhyme 'Ring-a-Ring o' Roses'. This describes the earliest sign of the pustules, which took the form of circular red rashes ('ring-a-ring o' roses'), the flowers that were used in an attempt to render the victims less harmful ('pocket full o' posies') and the sneezing, after which the sufferers die or, in the context of this rhyme, 'fall down'. Once a person had been infected, there was then a real possibility that the illness could be spread further in the home by the deadly bacteria which was contained in aerosol droplets exhaled from the lungs.

The Great Plague of London in 1665 was evidence of the continuing outbreaks which swept Europe in the centuries after the Black Death. It killed 70,000 people and was followed by the Great Fire of London in 1666, which is popularly believed to have eliminated the plague. Probably more significant were changes affecting the distribution of the black rat itself. It is certainly clear that the areas of London which were worst affected by the plague were not the same as those destroyed by the fire, as would have been expected if the flames were responsible for taming this infection. It is more likely that the black rat came under pressure from its heavier and more aggressive Asiatic relative, the brown rat. Although both species are capable of carrying bubonic plague, the life style of the brown rat is significantly different; it is primarily a burrowing species.

Even so, the Great Fire may have had some impact on hastening the black rat's decline. These rats favour wooden buildings of the type which were destroyed in the fire. At this stage in history, timber was the main material used in the construction of dwellings and commercial premises. Today, black rats still linger in areas where wooden buildings are commonplace, such as disused docks. Brown rats do not live in such close proximity to people and thus the risk of them spreading plague is correspondingly reduced.

It is not just in Europe and Asia that the brown rat has proved dominant. Further afield, in the USA for example, where the species was introduced to the east coast in about 1755, it has taken over whole areas, such as south-western Georgia, which were previously the exclusive domain of the black rat. In less than 150 years brown rats became established right across the USA.

Only in tropical parts of the world has the black rat been able to fare better than its relative, possibly because wooden buildings are still more widely used there than elsewhere.

There are few animals which have had such a dramatic effect on human history as the black rat. In England, the advent of the plague, especially the Black Death, caused widespread social disorder, such as the Peasants' Revolt of 1381, and, ultimately, change, as the established feudal system broke down because of such widespread loss of life.

More localized outbreaks of bubonic plague have continued until the present day and it took a long time to unravel the way in which the disease was spread. The first suggested link with rodents was documented by the author Daniel Defoe, best known for his novel *Robinson Crusoe*. Writing at the time of the Great Plague, he observed, in *A Journal of the Plague Year*, that not only dogs and cats were being killed in an attempt to stem the infection, but also mice and particularly rats were being poisoned in large numbers using arsenic bait.

Even when the third worldwide outbreak of bubonic plague began in the 1850s, the cause was still unknown. The effects of the infection then were just as devastating and tragic, however, as they had been in the course of previous centuries. Originating in China, the disease spread to India, where more than 11 million people died.

From the ports of the region, such as Hong Kong, the plague was transmitted around the world, thanks to infected rats. It reached Hawaii in 1899 and was confirmed on the mainland USA during the following year, which ironically was the Chinese Year of the Rat. The authorities in San Francisco were so horrified by the implications of plague that at the beginning they attempted to conceal the outbreak, in order to prevent mass panic. Worldwide, over 120 million people died in the succeeding 3 years.

Unfortunately, the delay in reacting to the plague enabled it to become established in the native rodent population. The infective fleas were acquired by the ground squirrels of this part of the western USA and although, in a bid to eliminate the infection, huge numbers were killed, this proved unsuccessful. In fact, ground squirrels were already probably the source of the bacteria which gave rise to the outbreak in Los Angeles in 1924. This was a particularly unpleasant pneumonic form of the disease. It was the last major outbreak of plague in the USA but the carriers of the disease are still present there.

The cause of bubonic plague, the bacterium *Yersinia pestis*, was finally identified in 1894 but it took a further 18 years before the key role played by rats and their fleas was fully appreciated. Plague also affects rats and fleas become infective by feeding on the blood of a sick rat. When the rat dies, the fleas will leave its body in search of other feeding opportunities. They may then bite people, thus transmitting the bacteria, which have been multiplying in their bodies, with deadly consequences. Ultimately, the fleas also die from plague, about 3 days after becoming infected. Outbreaks tend to subside in the winter in temperate areas because fleas are less active at this time of year.

Even today bubonic plague is a potential killer. Panic followed its emergence in India during 1994, with some countries banning aircraft and ships from this part of the world until the outbreak was under control. While antibiotics can help to overcome the bacterium and a vaccine is now available, this infection could spread around the world at a

Rodent-borne plague is still a constant menace today and a range of other zoonotic infections can be spread from rodents to people when they live in close proximity to each other.

frightening rate, thanks to modern means of transportation.

There are still reservoirs of plague linked with rodents on every continent today. This makes it very difficult to eliminate the disease. It is even possible for the infection to be spread a long way from the original source, particularly in areas where rodents are trapped for their fur. In Siberia, for example, marmots and susliks (Sciuridae) may suffer from the pneumonic form of plague and the cold climate of this region can allow infective fleas to survive for over a year without feeding.

Where rodents are living in close proximity to people, the risk of infection is increased. In parts of South America, for example, guinea pigs (*Cavia* species), which are kept as a source of food, roam in and around homes and can represent a major source of human infection.

In other areas of the world, native rodents may infect rats, which in turn then pass the illness on to people. In southern Africa, gerbils (Gerbillinae) can transmit plague in this way and, at the same time, the rodent population becomes dramatically reduced.

Rodent ticks can also represent a severe danger to human health, being capable of transmitting *Rickettsia* infections, which give rise to typhus. The ticks feed on the blood of an infected rodent and can then spread the illness if they attach themselves to a human being. Various forms of typhus are recognized, with Rocky Mountain tick fever, for example, being restricted to parts of North and South America. It causes a sudden fever and a characteristic rash that appears about 4 days after other symptoms, which usually include a severe headache.

This illness can be treated successfully with antibiotics but, if undiagnosed, can prove fatal. Ground squirrels (*Spermophilus* species) and voles (Microtinae) are a major reservoir of infection in North America.

Other complex cycles of infection involving rodents, invertebrates and people have also been identified. Blood-sucking sandflies (*Phlebotomus* species) are involved in the transmission of leishmaniasis, a protozoal infection. These flies consume not only blood, but also tissue cells in the dermal layer of the skin. If they feed on an infected rodent, they will ingest groups of protozoal parasites, called Leishman-Donovan (LD) bodies at this stage in their life cycle, which are present in the dermis. These parasites then escape from the dermal cells and start to multiply in the sandfly's gut. They start to move towards the biting mouthparts of the insect and some will be injected into the person or animal on which the sandfly next feeds.

There are a number of different strains of *Leishmania* around the world. In Asia, cutaneous leishmaniasis is also known as oriental sore, because this strain of the *Leishmania* parasite causes a localized ulcerating sore at the site where it is introduced into the body. Several rodents are the source of Chiclero's ulcer, another form of leishmaniasis, seen in forest workers of Mexico and British Honduras. This ulcer is often seen on the ear.

Control of leishmaniasis entails curbing the numbers of wild rodents and sandflies. The effects can be dramatic. In an area of Turkmenistan, the incidence of this infection in the human population was about 70 per cent before an extensive poisoning campaign of gerbils, which were

It is not just rats and mice which may harbour deadly diseases. Desert rodents, such as jirds (*Meriones* species), have been known to transmit infections, including plague, to rats, who have then passed these diseases to people.

Rodent droppings can spread a number of serious diseases, but it is not just their droppings which can be harmful. The urine of rats, for example, can transmit Weil's disease.

the reservoir hosts of the infection. More than half a million burrows were baited, wiping out a huge number of these rodents, and, less than a year later, there were hardly any cases of leishmaniasis to be found among the local people.

Although not often regarded as a source of rabies, rats can also carry this deadly virus. In a study carried out in Thailand, about 4 per cent of the rats tested were positive for this virus. It may possibly be spread through the rodent population by fighting. There are other viruses which may be hazardous to people in some parts of the world and which are spread by rodents.

The bacterial disease leptospirosis is a widespread hazard to other animals, such as dogs, as well as people. This rat-borne illness is linked closely to damp surroundings where rats are present, since the *Leptospira* bacterium survives best in such conditions.

The most serious form results from infection by *Leptospira icterohaem-orrhagiae*, which results in the human illness called Weil's disease. People such as sewage-workers and Australian sugar-cane farmers, working in localities where rats occur, and where there may be areas of standing water, are especially vulnerable. Bathing in canals and ponds in areas where there may be rats is equally hazardous. The *Leptospira* bacteria become concentrated in the rats' kidneys and are passed out of the body in the urine. These bacteria survive well in water and can enter the human body via any damaged skin, the mucous membranes, or the conjunctiva of the eye. They then cause infective jaundice.

Dogs that are used for ratting, such as terriers, can also be infected with a strain called *L. canicola*. They may spread this to their owners but the advent of a vaccine for dogs, which protects them from this type of leptospirosis, has led to a dramatic decline in the incidence of this disease, which can prove fatal for both dogs and people.

THE FUR TRADE

Some rodents have been particularly valued for the quality of their fur. In Europe, the beaver (*Castor fiber*) was almost exterminated for this reason, at a relatively early stage in history. In Britain, by the end of the twelfth century, the distribution of the beaver had been reduced to a small area of Wales. Elsewhere in Europe its numbers also plummeted because of hunting pressures. Beaver fur was in great demand, for both coats and hats.

Another source of supply for beaver fur was North America. French fishermen visiting Canada began to trade with the native Amerindians, returning home to Europe with furs. In 1599 this arrangement led to the development of the city of Quebec, around which the fur trade flourished.

The importance of beaver furs was such that, for a period, they served as the official currency of the colony. From the sixteenth century onwards beavers were also killed for their glandular secretion, called castoreum, which was believed to have valuable medicinal properties. In fact, this was true to an extent, since castoreum contains salicylic acid, which is the main ingredient of aspirin and has anti-inflammatory, analgesic and antipyretic properties, among others.

The trade led to the establishment of the Hudson Bay Company, under a Royal Charter from Charles II of England in 1670, as France and England vied with each other for the lucrative fur market.

The North-West Fur Trading Company was then set up in 1783. Trade switched from France to London and, although over 100,000 beaver skins were being traded each year, the peak occurred between 1860 and 1880, when more than 3.5 million pelts were sold. This level of trade was unsustainable and beaver populations across North America went into decline.

The change in fashion from beaver-fur to silk hats, at the end of the nineteenth century may have helped to save the depleted population. Since then conservation measures have been introduced and the numbers of beavers have started to increase, allowing a continued, albeit limited trade in their fur.

Another group of rodents brought to the point of extinction by the fur trade were the chinchillas (*Chinchilla* species) of South America. The Ancient Incas first hunted chinchillas and then, in the sixteenth century, the early Spanish explorers took the fur back to Europe from Peru. It was soon being used to decorate royal garments, because of its exquisite softness and rarity. Instead of having a single hair growing out of a hair follicle, chinchillas can have as many as 80, which gives them a very dense pelt. This protects them from the cold in the mountains of

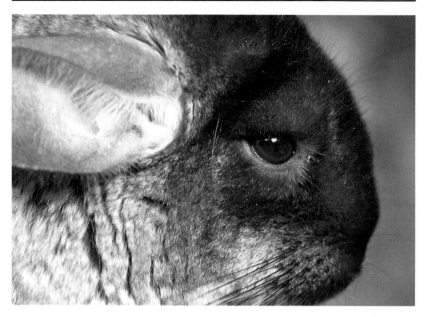

The soft, dense fur of chinchillas (*Chinchilla* species) almost proved to be their downfall, as they were heavily hunted, but captive breeding rescued them from extinction.

the high Andes where they live. Their coat is also so thick that parasites, such as fleas, cannot penetrate it.

Trade in chinchilla pelts became commonplace during the nineteenth century. From 1842, sometimes in excess of 200,000 skins were being sold annually in London by the Hudson Bay Company. Not surprisingly, these rodents had become almost extinct by the end of the century. It takes over 150 pelts to make a single full-length coat.

In South America early attempts at farming chinchillas to maintain the trade met with failure. Then, in the early 1920s, M. F. Chapman, a mining engineer, decided to try a different approach: adjusting the chinchillas gradually to the change in altitude, as they were brought down from the mountains nearer to sea level.

The fact that Chapman had to employ 23 men, over a period of 3 years, to obtain just 11 chinchillas is evidence of the endangered status of the chinchilla at that time.

Once they had become accustomed to captivity, these chinchillas were taken to California, where they started to breed freely. Until the 1960s, chinchillas were still bred almost exclusively for their fur but, since then, with the decline of the fur market, chinchillas have become popular as pets in many countries around the world.

Not all such introductions have been so successful, however. In fact, some have proved to be decidedly costly. There was a huge trade in coypus, or South American beavers (*Myocastor coypus*). This rodent is also known as the nutria, a term also used to described its pelt. The name comes from the Spanish word for otter.

Today, chinchillas are very popular simply as pets, and a wide range of colour variants have been bred as domestication has proceeded.

The trade in nutria pelts reached such a fever pitch in the early decades of the twentieth century, peaking at over 1 million in 1918, that the authorities in South America began to curb it. This in turn encouraged speculators in Europe to embark on farming coypus, which had first been bred commercially in France during the 1890s.

The boom of the late 1920s rekindled this industry for a time, with the French farms being concentrated mainly in the southwest and central areas of the country. When the market collapsed in the following decade, many of the remaining coypus were turned loose in the countryside. Here they bred and have continued to expand their range through the Loire valley and into the Camargue.

The first coypu farms in the UK were established in southern and southeastern England. At the peak of this enterprise there were 50 farms, but none continued after the start of World War 2 in 1939.

Although sightings of wild coypus were made in many areas, starting near Horsham in West Sussex in 1932, the major build-up in the UK occurred in the counties of Norfolk and Suffolk. Here the marshland and rivers were similar to their native habitat in South America.

Coypus grew in numbers until 1962, when the population reached a peak of about 200,000. By this stage they were having a marked impact on the environment. Both wild and cultivated plants were being destroyed in large quantities to feed the appetites of these coypus, which are one of the biggest species of rodent in the world today. Their burrowing habits were also damaging and the resulting collapse of river banks, leading to flooding, posed a threat to farmland.

A trapping campaign to curb the coypu was begun in the UK in 1962 and resulted in the capture of nearly 100,000 animals in just 2 years. This did not appear to have a noticeable impact on their numbers, although the weather proved helpful. The exceptionally cold winter of 1962/3 wiped out much of the population, confining them to an area of just 6,000 km² (3,750 sq. miles) in East Anglia, and a succession of prolonged cold spells served to keep the population in check.

Estimates of their numbers in the 1970s suggested a maximum of 14,000 individuals, in spite of continued trapping. The decision was then taken to push for the elimination of all coypus by the start of the 1990s. This target was subsequently met but elsewhere in many other European countries coypus are still present.

In the USA coypu farming became popular somewhat later than in Europe, beginning in earnest during the late 1930s. A small group of coypu were kept in the marshland around Avery Island where they thrived in semi-natural conditions. A hurricane struck the region in August 1940, however, allowing about 150 coypus to escape over the top of their enclosures when the marshes flooded. They spread quickly with reports of sightings being made from more than 100 km (63 miles) away barely a month later. These coypus multiplied rapidly and others were subsequently released within the state boundaries in the mistaken belief that they would help to keep waterways free from excessive plant growth.

While this may have been true to an extent, the coypus were not averse to turning their attentions to cultivated crops, particularly the rice growing in the adjoining paddy fields, and their burrowing habits also caused problems. Coypu are still tolerated in the USA, however, and are trapped both for nutria and as a source of meat. The scale of this enterprise now adds up to millions of dollars per annum.

Although in temperate areas, cold winters exert a major check on the numbers of coypu, elsewhere predation can be of greater significance. In Israel, for example, coypus were introduced because, unlike most other fur-bearing mammals, the quality of their fur is not adversely affected by the warm climate. Some escaped from the drainage canals of the Kafr Rupin and in the vicinity of the Naaman River. Their numbers are curbed by indigenous species, such as the jungle cat (*Felis chaus*), golden jackal (*Canis aureus*) and Egyptian mongoose (*Herpestes ichneumon*).

Coypus are also present in Africa but are not widespread, possibly because of the effects of predation. This was less of a controlling influence in the UK, where few species, with the notable exception of the red fox (*Vulpes vulpes*), represented any threat to them.

Another highly destructive rodent whose distribution has been extended widely as a result of the fur trade is the muskrat (*Ondatra zibethicus*). From being confined to North America, muskrats are now found in countries right across Europe and Asia, all the way to the shores of the Pacific.

The first fur farms for these rodents in the UK were established in the early 1920s. By the end of the decade there were more than 76 farms and escaped muskrats were becoming established in various parts of the country, particularly in the county of Shropshire. A determined campaign was then launched to eliminate muskrats from the UK and their farming was outlawed totally in 1932. Barely a decade later these destructive rodents had been exterminated, with nearly 4,000 having been trapped in the intervening years.

A key difference between the muskrat and the coypu is the fact that the muskrat can survive successfully on the northern shores of Canada and is well adapted to withstand prolonged periods of cold weather. Not surprisingly, therefore, this pattern has been repeated in the Old World, where muskrat populations are established in many northern areas, such as Sweden, where the water is actually frozen for about two-thirds of the year.

Muskrats are also well established in adjacent northern parts of the former USSR. They have crossed from there into Mongolia, where they were first recorded during the 1940s, and then into China. The Chinese population of muskrats has now built up to such an extent that more than 0.5 million pelts are now being traded annually.

The spread of the muskrat in little more than 50 years reflects the tremendous adaptability of this species. Like the coypu, however, their presence in the landscape is not entirely benign. Their burrowing habits can result in flooding and, because they share a habitat with rats, they now also pose a potential health hazard to people, being carriers of Weil's disease in some areas. Muskrats will also feed on cultivated crops, including sugar beet, causing serious economic losses when they are numerous.

These rodents face few predators through much of their extended range but, in certain localities, they have built up to such numbers that they have exceeded the available food supply. Consequently the populations in these areas have fallen sharply.

With a population of muskrats now established in the wild in Argentina and Chile, from stock released during the 1940s, this species now occurs over a wider range than any other introduced mammal. The case of the muskrat provides further confirmation of just how adaptable certain rodents can prove when presented with an opportunity to extend their distribution.

Not all species can manage to switch between continents with such ease as the muskrat, however, as exemplified by the attempts made to introduce the North American beaver (*Castor canadensis*) to both Europe and Asia. In the UK these beavers were kept at three localities, on the Isle of Bute in Scotland, in Suffolk and in Sussex but, although small populations survived for a number of years, they never thrived and ultimately died out. A release in France was equally unsuccessful.

North American beavers have now been introduced to Finland and are currently more numerous than the endemic European species (*C. fiber*), although they sometimes live alongside each other. Releases here began in the late 1930s and, today, these beavers are most numerous in the east of the country. However, in a region where the forestry industry is of considerable economic importance, they are not universally popular, because of the damage which their logging and damming activities cause to the trees. Such losses far outweigh the value of the beaver pelts, which are only harvested in small numbers.

THE INTRODUCTION OF ALIEN SPECIES

The potentially harmful effects of introducing alien species to new environments are clearly appreciated today and many countries have legislation which outlaws the release of non-native species to the wild. In Victorian times, however, such introductions were often positively encouraged, with the result, for example, that rabbits (*Oryctolagus cuniculus*) and red foxes (*Vulpes vulpes*) were taken to Australia and set free.

In the UK, grey squirrels (*Sciurus carolinensis*) from North America and edible dormice (*Glis glis*), originating from the European mainland, were liberated as part of the vogue for bringing new creatures into the country.

Early attempts to establish grey squirrels in the UK were made in Wales during the 1820s but the major period of releases was in the first three decades of the twentieth century. Those which had been liberated previously by the Duke of Bedford, at his estate at Woburn Abbey, Bedfordshire, were already thriving and some of these squirrels were moved to other localities.

By 1930, the distribution of grey squirrels extended over the southeast of England as far as the counties of Warwickshire and Northamptonshire and covered an area of about 25,000 km² (9,650 sq. miles). Today, they can be found throughout England and Wales, wherever there is suitable habitat. Another population exists in the lowlands of Scotland. In 1913, grey squirrels were also introduced to Ireland by the Earl of Granard, at Castle Forbes in County Longford, where they also became established.

Grey squirrels have not been taken to other European countries but they were introduced to South Africa, originally by the former Prime Minister, Cecil Rhodes, in the early 1900s. He released them on his estate at Groote Schuur but before long they had spread from there into the adjoining area of Cape Town. They are now particularly common wherever there are oaks and pine trees, which provide food for them. These squirrels are also likely to be found in agricultural areas, where fruit-growing predominates, although they have not colonized areas of grazing land, where the habitat is unsuitable.

Rather strangely, however, grey squirrels appear to cause less damage to the trees in South Africa than they do in the UK. This could be related to the wider choice of food available in the UK. Predators of grey squirrels in South Africa include the boomslang snake (*Dispholidus typus*), which hunts them in the trees, moving with great speed before inflicting a deadly bite.

For a period grey squirrels were introduced in the vicinity of Melbourne,

Some wild rodents can become quite tame, particularly squirrels, such as the grey (*Sciurus carolinensis*), which are diurnal and live in close proximity to people, in parks and other areas where they are not persecuted.

The grey squirrel (*Sciurus carolinensis*) causes damage in both forestry plantations and urban areas. It is not unknown for them to invade houses, usually through the roof space, and, like rats, it may sometimes bite through electrical cabling, with serious consequences.

The grey squirrel (*Sciurus carolinensis*) is not native to Europe but was first introduced to the UK from its North American homeland in the 1820s and is now well established. It was also taken to South Africa.

Australia. Although they appear to have bred, conditions were clearly not entirely satisfactory and the populations have now died out.

The northern India palm squirrel (*Funambulus pennantii*) has found conditions around Perth in Western Australia more to its liking. A small group was set free in the grounds of the South Perth Zoological Gardens at the end of the last century. Today, they have spread into some of the residential areas, particularly in Como and southern Perth, but the open countryside around Perth has probably served to restrict their spread. These squirrels appear to cause little if any damage.

Also in the southern hemisphere, attempts made in 1869 by the Auckland Acclimatisation Society to establish free-living guinea pigs in New Zealand ended in failure, although a colony has become established on the island of Santa Cruz, part of the Galapagos group.

THE IMPACT OF ESCAPEES

The successful establishment of populations of muskrats, grey squirrels and other introduced rodents in parts of Europe and elsewhere has been due largely to the human protection which they received in the critical early stages. Where only a few animals were involved, and where they were hunted soon after their escape, colonies have failed to become established for any length of time. For example, a small breed-

Hystrix porcupines are believed not to be native to Europe. However, there is now a population in southern Europe and escapees from a zoological collection lived wild and apparently bred for a period in the UK, before being eliminated in the early 1970s.

ing group of Himalayan porcupines (*Hystrix hodgsoni*), stemming from a pair which escaped from a wildlife park, became established in Devon, UK. As a result of the serious damage which they caused to the conifer plantations in the area, a trapping campaign was begun which eliminated these rodents before they could spread further afield.

RODENTS AS FOOD

A wide range of rodents are hunted for food, a practice which dates back centuries. The Romans regarded the edible dormouse (*Glis glis*) as a great delicacy in the autumn months when their bodies were heavy with fat to sustain them through their period of winter dormancy. The Romans even fattened these rodents specially in clay pots, or 'glisaries', feeding them on acorns and chestnuts, or in enclosed areas outdoors where suitable nut trees were growing. Dormice are still eaten in France and other parts of Europe today.

In South America, the guinea pig has been domesticated for over 2,000 years. These rodents were already being widely bred by the Incas at the time of the Spanish invasion in the late sixteenth century. Colour variants were then in existence and these guinea pigs were kept in and around the homes.

As well as providing a source of food, they played a part in the religious ceremonies of the Incas, frequently being mummified, rather like cats in Ancient Egypt, although their precise role is unclear. Although the Brazilian species of cavy (*Cavia aperea*) was once thought to be the original ancestor of the guinea pig, because of the crest of hair on its neck, it is now generally accepted that the Peruvian cavy (*C. tschudii*) is more likely to be the ancestral form. As many as 7 million guinea pigs are eaten each year in Peru.

Rats are hunted both in parts of Asia and Africa as sources of food and attempts are now being made to farm the giant pouched rat (*Cricetomys gambianus*) in west Africa. Remarkably, captive breeding of these rodents has revealed that, in just five generations, they lose all fear of people and this increases their farming potential. Adult males can weigh up to 1.5 kg (3¼ lb), while female giant pouched rats are slightly lighter. They have a high reproductive rate, with females being able to produce as many as ten litters a year, each comprising about four pups on average.

Other species of rat are highly valued as a source of food. In Ghana, cane rats (*Thryonomys swinderianus*), known locally as 'grasscutters' can sell for nearly double the price of meat from domesticated animals, such as pigs. The meat of these rats is usually smoked and estimates of the trade show that as much as 200 tons may be eaten annually in Ghana alone. This is equivalent to more than 41,000 rats!

RODENTS AS PETS

In Victorian times rats in the UK were particular pests, whose numbers had to be curbed in the rapidly growing towns of the Industrial Revolution. Even so, there was money to be gained by using live rats for entertainment purposes. In the public houses of this period there were often rat pits, where dogs, encouraged by their owners and a host of spectators, would be expected to kill as many rats as possible in a set period of time. Some were highly effective rodent-killers, with one terrier, called Billy, managing to dispatch a record 500 rats in just 5½ minutes.

A large supply of rats was clearly needed, with one publican employing 20 rat-catchers for this purpose. They would catch the rats free of charge and then sell them to the landlords, but generally those from sewers were avoided because they were most likely to make the dogs ill.

Among the thousands of rats which were caught, the rat-catchers sometimes encountered strangely coloured individuals. These were often kept alive as they were regarded as something of a novelty. Jimmy Shaw, who ran one of London's most famous rat pits, owned a collection of such rats. His first white rat had been caught in a cemetery and he also acquired black-and-white rats from various localities.

Another man with a professional interest in rats at this stage was Jack Black, who was Queen Victoria's official rat- and mole-catcher. He claimed to have obtained and bred pied rats with great success and sold them around the country. They were popular pets apparently, especially with young women, who often kept them in squirrel cages.

Beatrix Potter was probably one of the first female rat-keepers, doting on her pet rat, which she called Sammy. In 1908, when her story *Samuel Whiskers* was published, Beatrix dedicated it to her pet. The popularity

Rat-trapping was a major enterprise in Victorian London. The rats could be sold alive to supply the rat pits of public houses, where dogs killed them in timed contests. Rats from underground burrows rather than the sewers were preferred because they were reputedly healthier.

of rats as pets seemed to be assured. They started to be shown and breeders began to develop strains with new colours and patterns.

Interest had faded by the late 1920s and, although brief revivals occurred during the late 1950s and again in the 1960s, it was not until the mid-1970s that rat keeping became a popular hobby. This is probably a reflection of changing attitudes. People today in North America and Europe are less likely to have direct contact with wild rats than their ancestors and the fear associated with these rodents has been lessened as a result.

Approximately 25 distinct varieties of fancy rat now exist. All are descended from the brown rat (*Rattus norvegicus*), although for a time during the early 1920s, several colour variants of the black rat (*R. rattus*) were being bred. The most unusual of these was undoubtedly a bizarre greenish-coloured strain.

Rats have also become established laboratory subjects, being specially bred for this purpose. About 20 million rats are used each year for research purposes in laboratories in the USA. Medical research, nutritional investigations and psychological studies – rats have been used in all these fields. They have also been sent into space in the name of science, starting in 1960 when a pair returned successfully to earth after being sent into orbit as part of the Russian space programme.

The image of the plague-infected rat is far removed from that of the rats used in laboratories today. In 1956 the production of pathogen-free rats was pioneered at the Charles River Breeding Laboratories, based in Wilmington, Massachusetts, USA. Such rats can be vital for tests where standardization is essential and they obviously need to be housed in special facilities.

Today's strains of domestic mice have a similar history to that of rats. Many of the colour varieties seen today were originally evolved in Victorian laboratories. There are now more than 700 possible colour and coat variants, with fancy mice ranging in colour from pure white through fawn and red to black. Colour combinations such as black-and-tan are also well established. Long-haired, short-haired and sleek-coated satin strains are all in existence. Hundreds of thousands of fancy mice are kept today, either as pets or for exhibition purposes, and judging standards are now established for the different varieties.

Surprisingly, the ancestry of today's tame mice may date back thousands of years, as shown by studies on their genetic material, DNA (deoxyribonucleic acid), which have been carried out at the University of California. The original mice which gave rise to many of the laboratory strains can be linked back to Ancient Greece, where, on Tenedos Island, a temple was built to honour Apollo, who was considered to be the god of mice. This temple commemorated a famous battle in which the mice destroyed the bows of the enemy archers. For generations, the priests at the temple had continued to breed these tame white mice and from here their descendants may have spread around the globe.

Today, mice are used in even greater numbers than rats for laboratory research. Special strains with particular susceptibilities are favoured for particular areas of investigation, e.g. the so-called 'nude' mice, which

Colour varieties are common in domestic rodents such as this Mongolian gerbil (*Meriones unguiculatus*) – proof of their genetic adaptability.

lack hair, are used for investigations into skin conditions. Laboratory mice were also the first animals on earth exposed to moon dust in order to assess whether it contained any harmful microbes.

The term 'guinea pig' has, of course, acquired a new meaning because of the animal's value in testing pharmaceutical products in particular. These rodents are also widely kept as pets and for exhibition.

There is some dispute about the origin of their name. Perhaps, during the sixteenth century, when they were first brought to England, the guinea pig sold for the sum of a guinea, as well as having a distinctly porcine outline. Alternatively, ships plying the Atlantic crossed with the assistance of the trade winds and so frequently docked in West Africa before returning northwards to Europe. The mistaken belief that these rodents came from Guinea, rather than Guiana, may have led to them being called guinea pigs. Also they often grunt like pigs when touched. Like other pet rodents, the variety of colour forms and coat types of guinea pigs has increased greatly since they became popular pets.

In some cases, it seems merely fortuituous that a rodent has developed into a widely kept species. The Syrian hamster (*Mesocricetus auratus*), for example, would probably still be a little-known desert-dwelling rodent if it were not for Professor Saul Adler, who was based at the Hebrew University of Jerusalem.

Guinea pigs are the most popular pet rodents today, and justifiably so, as they can be handled easily, even by children, and are most unlikely to bite. They are bred in a wide range of colours and coat types.

31

He required hamsters to use in his studies on the parasitic blood disease, leishmaniasis. A colleague of his, Israel Aharoni, from the University's zoology department, was recruited to help with this quest. He managed to trap a female hamster with 11 pups in a wheatfield near Aleppo, in Syria, in a burrow which extended for about 2 m (6½ ft) below the surface.

One of the offspring was killed by its mother, so Aharoni and his wife were forced to hand-rear the survivors. Another young hamster escaped but worse was to follow. Back at the University, the remaining hamsters were placed in the animal house, in a cage with a wooden floor.

This proved disastrous: they gnawed their way out overnight and five of them disappeared. The remaining four then provided the breeding nucleus from which virtually all of today's pet hamsters around the world are descended. They proved to be prolific, producing 150 youngsters within a year.

Adler soon appreciated the value of these hamsters in his work and smuggled two pairs into the UK in 1931. He passed them to a colleague, Professor Edward Hindle, who also discovered their prolific breeding habits. Hindle donated some of the hamsters to London Zoo in the following year, and the Zoo was soon over-run with young hamsters.

In 1937 the Zoo transferred some surplus stock to private keepers for the first time, and so began the keeping of pet hamsters. It was largely due to the vision and enthusiasm of Percy Parslow, however, that hamsters built up such a strong following among the pet-seeking public. He established the UK's first hamster farm at Great Bookham, near Leatherhead, in Surrey. Today, hamsters are still being bred commercially there to supply the pet trade.

For more than 40 years the entire captive population of Syrian hamsters around the world was descended directly from the original four youngsters that were hand-reared by Aharoni. There appears to have been no adverse effects from the necessary in-breeding, which took place over probably more than 100 generations down through the intervening years.

It was not until 1971 that any more wild hamsters were obtained. Thirteen were taken to the USA in that year and several imports have since occurred, introducing new genes to the established stock.

Over 30 colour varieties of the Syrian hamster have now been developed and coat variants are also common. In recent years, however, the dominant position of this species as a pet has been challenged by others, whose development was also encouraged by Percy Parslow.

Russian dwarf hamsters (*Phodopus* species) were largely unknown to science until about 1961, when they were first kept in the former USSR by the Leningrad Zoological Society. Subsequently stock was sent to London Zoo, where these hamsters were first bred in 1968. It was at this stage that Percy Parslow obtained some of these hamsters, which are significantly smaller than the Syrian species, and began breeding them.

The Chinese hamster (*Cricetulus griseus*) is another species which is quite well established among private breeders. Their ancestry can be

The Syrian hamster (*Mesocricetus auratus*) has a tiny area of distribution. It became a popular pet by chance, and is better known as the golden hamster.

traced back to a small number obtained from just outside the Chinese capital of Beijing in 1925.

Another species of rodent which has built up a very strong following as a pet is the Mongolian gerbil (*Meriones unguiculatus*). The famous French missionary, Père David, was the first Western zoologist to encounter these gerbils, in the 1860s. But it was not until 90 years later than some Mongolian gerbils were taken abroad for the first time, to Japan. Here they started to breed readily and stock was then sent to the USA, where at first they were also kept for laboratory purposes.

Before long the engaging nature of these rodents placed them in great demand among pet-seekers, and so began their rapid rise to their current high level of popularity. A host of colour variants, including pink-eyed whites, shades of cinnamon, lilac and almost totally black Mongolian gerbils are now well established.

This is unlikely to be the end of the list as far as rodents which have been kept for other purposes and then become pets is concerned. Chinchillas are increasingly being kept as pets, rather than for their fur, and chipmunks are also proving popular as companion animals.

Colonies of spiny mice (*Acomys* species) are being bred in increasing numbers by rodent enthusiasts and so are likely to be more widely seen as pets in the future.

Human curiosity will result in other rodents joining this list of those which have been welcomed into our homes. Spiny mice (*Acomys* species), with their peculiar coats and breeding behaviour, are a source of fascination to specialist rodent-keepers at present, as are the multi-mammate mice (*Mastomys* species). Gundis (*Ctenodactylus* species) and other members of the cavy clan may also soon be added to this list.

There is also the real possibility that captive breeding of rare rodents could result in surplus stock becoming available to dedicated private breeders, and possibly even to the pet trade, once their numbers have been increased. Most of the rodents under threat today, like the hutias (*Capromys* species), are endangered because they have a relatively small area of distribution, which is undergoing habitat change. They may also be subjected to human persecution.

There has been considerable success in breeding the Jamaican hutia (*Geocapromys brownii*) at the Jersey Wildlife Preservation Trust. This has enabled over 40 of these rodents to be released back into the wild. Other species are also breeding in various collections, often in the capable hands of enthusiasts, and this gives hope for the future.

RODENTS IN MYTH AND LEGEND

One of the most enduring figures linked with plagues of rodents is the Pied Piper of Hamelin. The origins of this legend can be traced back far beyond the Middle Ages, however, drawing on Sanskrit mythology.

Popular legend tells how, in 1284, when Hameln, in Germany, was being plagued by rats, a piper volunteered to free the town from this curse. No one knew his name but, since he dressed in a costume of two colours, the man was called the Pied Piper. True to his word, he lured the rats out of the town with his pipe and they drowned in the Wessex River.

When the people refused to pay him, the mysterious Pied Piper disappeared, returning on 26 June to pipe away the 130 children of the town, apart from two who were blind and lame. He led them into a cave on Koppelberg Hill and they were never seen again. Interestingly, the story was modified later, describing how the children were actually taken around the mountain and subsequently lived in Transylvania.

As to how much, if any, of the Pied Piper story in its many forms is based on fact is impossible to determine. The modern version is drawn from the poem 'The Pied Piper of Hamelin', written by Robert Browning, for a boy called Willie MacReady, who was the son of a friend. Browning never intended to publish these verses but now the poem is probably his best-remembered piece of work.

Today in the medieval town of Hameln, it is prohibited to play music through Bunger Street, where the Pied Piper is reputed to have led the children away. Nevertheless, rats remain a common sight in the town, with rat-like images being well represented in the shops, for sale to eager tourists.

Similar, more recent tales exist elsewhere. In the town of Isfahan, for example, in the late seventeenth century, the people agreed to pay a dwarf called Giouf to free them of its rats in an hour. But having piped the rats from the town into the nearby Zenderou River, Giouf received counterfeit coins for his trouble and so he apparently returned to take revenge on the people.

Not all portrayals of rats have been menacing, however. The water rat in Kenneth Grahame's *Wind in the Willows*, for example, is a friendly, dependable creature. Nevertheless there is a certain fascination with huge rats, such as those mentioned in another classic, *Gulliver's Travels*. Other rodents have enjoyed a more sympathetic portrayal, such as the dormouse in Lewis Carroll's *Alice in Wonderland*.

In the world of cinema, while many films of the horror genre feature rodents, especially rats, cartoons often show them in a more sympathetic light. The natural cunning of rodents was portrayed in the popular long-running *Tom and Jerry* cartoons, with the mouse invariably outwitting the cat.

But undoubtedly the most enduring image of a lovable cartoon rodent is Walt Disney's creation of Mickey Mouse, who first came into existence in 1923. Disney himself had fallen under the charm of the mice which appeared in his room in Kansas City, Missouri, where he used to work late at night. His favourite was a bold individual which he christened 'Mortimer'. When he left Kansas City for Hollywood, it is said that Disney liberated the mouse in a field.

His interest in mice, however, had begun much earlier, while he was still a boy. The young Walt Disney was entranced by a cartoon strip called *Johnny Mouse*, drawn by a man called Clifton Meek. This encouraged Disney to start drawing mice on the walls of the barn at home.

The image of mice, and specifically Mortimer, remained in Disney's mind. It was while relaxing on a train journey across the USA, on 16 March 1928, that Walt Disney first drew the image that has since become internationally recognized across the world. Mickey Mouse took shape, complete with his familiar red outfit, yellow shoes and white gloves, between the towns of Toluca in Illinois and La Junta in Colorado. Although Disney's own inclination was to call his creation Mortimer Mouse, his wife Lillian persuaded him that Mickey was a better name. It was also at her suggestion that Disney created Minnie Mouse, and so two stars were born. In fact, much of the subsequent success of the Disney empire was thanks to Mickey and Minnie Mouse. The main difference from those early drawings was that Mickey ultimately lost his tail.

Even so, it took time for Mickey to gain public acclaim. His first appearance was in a silent cartoon called *Plane Crazy*, inspired by the transatlantic flight of the pilot Charles Lindbergh. The sudden surge of interest in talking movies, however, presented a challenge to Disney and his small team. With little money, but a lot of determination, they converted the Mickey Mouse epic, *Steamboat Willie*, into the world's first cartoon movie with sound, after the silent version had been rejected by the distributors. It opened at the Colony Theater on Broadway on 19 September 1928 to immediate acclaim, launching Mickey Mouse to international stardom.

Subsequent generations of children have grown up with Mickey Mouse as part of their lives. His image has appeared on more than 5,000 different commercial items and his name was even taken as the codeword for the Allied invasion of Normandy on 6 June 1944. It is fair to say that no other mouse has had such a great impact on history.

CONTROLLING RODENTS

Under favourable conditions, the populations of some rodents can build up to huge numbers in a very short space of time. A single pair of brown rats (*Rattus norvegicus*) could potentially have 15,000 descendants in a year. Natural levels of predation are such, however, that female rats manage to produce an average of only 20 surviving youngsters during this period. Almost all wild-born brown rats have a life expectancy of less than a year.

Even so, they will battle ferociously for their lives. There are numerous cases of rats attacking dogs, cats, and even people when they are cornered, launching a pre-emptive strike against their opponent. Rats will also attack babies, and even kill them, and rat-bite deaths still occur virtually every year in the USA.

In the battle to control rats, some dogs have been specially bred for this purpose. The terriers in particular possess both fearless tenacity and the speed to dispatch the individual rats quickly.

Probably the earliest poison used in the battle against rats was red squill. Its potency for rats had already been appreciated over 3,500 years ago. It is derived from a plant (*Squilla maritima*) which grows in the Mediterranean region and has some medicinal properties, including an emetic action. If people inadvertently consume red squill, it makes

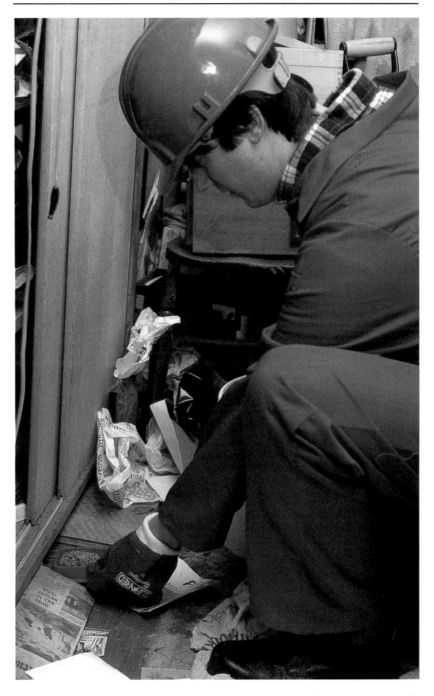

The use of poisons in the battle against rodents requires constant vigilance and new types of poison are constantly being required as rodent populations develop immunity. Care that poisons do not kill other animals, especially if they eat the poisoned rodents, is an important consideration for scientists today.

them vomit, but rats are unable to respond in this way. Instead, the poison remains in their bodies and paralyses the heart muscle. It is most effective against brown rats simply because they will take this bait more readily than their black relatives (*R. rattus*).

Rats are very suspicious when it comes to bait and, in a colony, only one member will try any unfamiliar foodstuff. Should this subsequently prove to be lethal, it can be almost impossible to persuade the other rats to take the bait. For this reason, red squill is not ideal, although it does kill rats very effectively if they can be persuaded to take it. Moreover, it represents no serious danger to people.

In the early days of toxicology, the aim was to make compounds which were poisonous to rats at a single dose in order to avoid this problem of bait avoidance. However, substances of this toxicity could also be lethal to other species, and dogs, for example, could die after consuming poisoned rats. One compound, sodium fluoroacetate, is so deadly that just a fraction of a gram can kill a human being, let alone a rat. This type of bait was not suitable for uncontrolled use and so research was undertaken to develop a slower-acting poison that represented less of a danger to human health.

In 1950, scientists investigating the deaths of cattle which had eaten sweet clover hay found that a certain chemical, called dicoumarol, was responsible for the fatal haemorrhages seen in their bodies. Testing this on laboratory rats, the scientists found that dicoumarol was invariably fatal to the rodents as well. It caused them to suffer massive internal haemorrhages. Further study at the laboratories of the University of Wisconsin, USA, revealed that only minute doses of the chemical were needed and that rats would not avoid dicoumarol in the same way that they avoided other existing poisons.

Launched originally under the name of warfarin, a host of similar products have subsequently become available. Some are in the form of a powder which the rat picks up on its coat and ingests when grooming itself. A single dose alone is not deadly, which is a safety factor of these chemicals, should children gain access to them.

As might be expected, it was only a matter of time before rats started to become resistant to these warfarin baits and were able to consume large quantities with no apparent ill effects. The first positive case of warfarin resistance was confirmed in Scotland in 1958 and similar findings were soon reported from other areas where this type of poison had been used.

The incidence of resistance varied widely, with up to three-quarters of the rats tested in Chicago being positive. Warfarin acts by blocking the action of Vitamin K, which is involved in the production of the clotting factors that arrest bleeding. Under normal circumstances, if a rat ingests a high dose of anticoagulant, Vitamin K stimulates the production of these clotting factors. If the rat has fed on warfarin, the level of clotting factors in the blood declines and so any blood loss continues unchecked.

The so-called 'super rats', with resistance to warfarin, soon become

Effective control of rats and other commensal rodents depends to a large extent on improved hygiene, both in and around human dwellings, especially in city areas.

dominant as the non-resistant rats in the population are wiped out. This is where the rat's tremendous reproductive potential comes to the fore, enabling numbers to build up again rapidly.

The development of this resistance has led to the formulation of new poisons to curb these rodents, but it is unlikely that anything other than an equilibrium will ever be established in the battle against rats. Their natural reluctance to sample unfamiliar foodstuffs means that persuading them to take the bait can be difficult.

Gassing rats in their burrows can be equally unsatisfactory; unless the gassing is carried out by experienced operatives, the rats simply slip out through one of a number of entrances before the gas overwhelms them. The toxicity of the gases used may represent a hazard in urban areas, near buildings, although gassing is accepted as the most effective method of dealing rapidly with a high concentration of rodents.

Trapping can only deal with a relatively small number of rats, and, again, because of their natural intelligence they soon come to recognize the danger posed by traps.

Ironically, the number of rats in cities around the world is probably increasing again because of the careless habits of human beings. We are currently creating a better world for rats with our throw-away life styles. There are rich pickings to be had for rats on the streets and plenty of suitable retreats for them. This is now being recognized by those who

Poisons are generally slow acting rather than being immediately lethal. This prevents rodents from becoming suspicious about baits. Traps can also be used to catch rodents but this is a relatively inefficient method compared with poisoning or gassing.

have to battle with these rodents. The most effective means of controlling rat populations is to make the environment less hospitable for them. Reducing their food supply through improved hygiene is one approach. Today's tendency to leave garbage, including the remains of food, on the streets in plastic bags, rather than in metal or plastic bins, is just one example of how rodents can thrive on human hand-outs. The high reproductive rate of rats means that they are ideally placed to exploit a favourable shift in environmental conditions.

Mice are generally easier to eliminate than rats because they are often less suspicious of poison and traps. Anticoagulant rodenticides have proved to be equally effective against mice but, again, resistance to such poisons has now been reported, with the first definite case being confirmed in 1960.

Second-generation poisons of this type proved highly potent for a period against these so-called 'super mice' but, as with rats, there are now mice which have developed immunity both to bromadiolene, for example, and to the early anticoagulants, such as warfarin.

A new means of deterring rodents from entering domestic buildings has been the use of ultrasonic devices. The high-frequency sounds generated, which are above the range of human hearing, reputedly cause the rodents discomfort and so they leave the area. Tests regarding the efficacy of this method of rodent control have given very mixed results. This may be partly because the sounds are not especially effective at penetrating the confined spaces where rodents tend to spend much of their time.

It is not just in urban areas, however, where rodents can be major pests. Crops can attract their attention at any stage from sowing onwards. Advances in agriculture have not always helped to reduce losses caused by rodents. The advent of pelleted seeds, which makes the sowing process easier, also make it simpler for mice to locate the seeds in the ground. Previously, the small size of many seeds meant that a significant proportion would escape the attention of rodents, particularly when they were planted at an adequate depth below the surface.

The use of poisons in the agricultural environment is clearly dangerous, especially for non-target species. But there is also a risk that ailing poisoned rodents could fall victim to predatory species, in whose bodies the poison may accumulate. This may well be the reason for the decline of the barn owl (*Tyto alba*) in some areas of the UK.

In the UK, the wood mouse (*Apodemus sylvaticus*) is the major threat to crops in fields, rather than the house mouse (*Mus musculus domesticus*), which is closely linked with human settlements. In contrast, in eastern Europe, the northern or field house mouse (*M. m. musculus*) is a more free-living variant, at home both in buildings and in the fields. There is a markedly obvious narrow line separating these two subspecies in central Europe. It reflects how the species has evolved along separate lines as its members have spread further westwards from Asia across Europe, and creates difficulties in controlling their numbers.

Form and Function

DENTITION

Various factors have helped rodents to become the numerically dominant group of mammals on this planet. One of the most significant features overall is their dentition, which has enabled them to utilize sources of food that other mammals are not equipped to deal with and which may also be used to help them seek shelter.

In spite of their diversity in size and appearance, all rodents have the same pattern of teeth and all have a pair of relatively large incisors in both the upper and lower jaws. For some time, this feature led rodents to be classified with rabbits and hares (order *Lagomorpha*), but lagomorphs have a second, much smaller pair of incisors alongside their main incisors, as well as other distinctive differences. Nevertheless, it is clear that these two orders share a common ancestry, with rodents and lagomorphs having diverged at some stage during the Palaeocene epoch, about 60 million years ago.

The majority of rodents have 22 teeth or less; the Australian water rat (*Hydromys chrysogaster*), for example, possesses just 12, which may be related to its predatory nature. At the other end of the scale, the African silvery mole rat (*Tachyoryctes argenteocinereus*) has 28 teeth.

The power of the incisor teeth in action can be seen in the case of this Cuban hutia (*Capromys* species), biting through a stem.

Whole leaves and stems will be eaten, as shown by this guinea pig (*Cavia porcellus*), which is finishing off eating a dandelion leaf.

The incisors of rodents correspond to the second incisors found in the typical mammalian pattern of dentition. Without them, rodents would encounter great difficulty in eating and so these milk teeth persist throughout the animal's life.

The roots of the incisors, located well back in the mouth to give good anchorage, typically extend as far as the level of the molars in the lower jaw. In the case of the mole rats, which rely on these teeth for tunnelling purposes as well, the roots of the lower molars are anchored behind the point at which the jaws articulate. This arrangement provides greater strength.

The upper incisors are more curved than the lower ones and both sets meet at a point. When viewed in cross-section, this point is at the front of the teeth, where a cutting edge is formed. Since the teeth must match up, the loss of one will result in the corresponding tooth in the other jaw growing in an abnormally spiral fashion, seriously compromising the rodent's ability to eat.

The shape of the tooth depends on the wear imposed upon it and, in rodents, the coating is significant. Hard-wearing enamel is generally present on the front of the incisors, with softer dentine behind. This arrangement contributes to the sharpness of the cutting edge.

The structure of the enamel covering can also be significant, both in tracing the evolution of rodents and in assessing the possible relationships between individual groups. The inner layer of enamel is composed of prisms which cross each other and so provide support, preventing the enamel coating from cracking. Above, there is a layer of radial enamel.

The most basic of the enamel types which have been identified is the form known as pauci-serial. This is not found in living rodents – only in their extinct ancestors. The number of layers in the prism bands within the teeth in this case is between two and four. Among living rodents today, the multi-serial form, where more than four prisms are present in a band, is regarded as being less evolved than the uni-serial, in which each band corresponds to a single prism and, being narrow, provides the greatest strength. The teeth themselves are often whitish in colour but, in some species, may have an orange hue.

The function of the incisors is to gnaw and the name 'rodent' is derived from the Latin word *rodere*, which means 'to gnaw'. Rodents have no canine teeth and there is a gap, the diastema, between the incisors and the cheek teeth. This enables the rodent to gnaw with its incisors while keeping the sides of its cheeks drawn into the mouth behind. As a result, unwanted material cannot enter, allowing the rodent to ingest only the food, and not contaminants, such as soil.

Early in their evolutionary history, some rodents had up to two pairs of premolar teeth in their upper jaw and a single pair of premolars in their lower jaw. Today, in many rodents, the premolars are either lacking entirely or are relatively inconspicuous, as in the mountain beaver (*Aplodontia rufa*), which is believed to be one of the most primitive of

This view of the underside of the upper part of the skull of a brown rat (*Rattus norvegicus*) shows the sharp incisor teeth at the front of the jaws, with a gap behind to the cheek teeth.

Many rodents, such as this grey squirrel (*Sciurus carolinensis*), use their front paws to hold food. Sitting in this position also enables the squirrel to detect danger more easily as it feeds in the open. If disturbed, it will drop its food and scamper up a tree.

today's rodents. Rats, mice and related families, including dormice, have no premolars, whereas other groups have some premolar teeth present in their jaws. Members of the squirrel family (Sciuridae) typically have a pair of small premolars in the upper jaw when they are young, but these milk teeth are shed and not replaced as they grow older, reducing the number of teeth from 22 to 20 at maturity.

The molar teeth are vital in all rodent species, enabling food to be ground up so that it can be swallowed easily. There are typically three pairs of molars in each jaw although, in Shaw-Mayer's mouse (*Mayermys ellermani*), from the highlands of northeastern New Guinea, there is just a single pair in each jaw.

The molar teeth of rodents are variable in appearance and this is used to distinguish between the different groups for classificatory purposes. Folds of enamel often run through these teeth, strengthening them.

The relatively rectangular shape of the molar teeth of squirrels reveals their long evolutionary history. The roots are deep but the molars themselves only protrude a relatively short way above the gum line.

Having taller molars has two advantages: it enables a rodent to grind up tougher particles of food and also means that the teeth themselves should suffer less damage. Refinements such as humps, sometimes linked together by bridges, serve to concentrate the grinding power of the molars.

45

Enamel not only coats the outside of the tooth, but also directly contributes to the action of the molars. It is abraded more slowly than dentine, which makes up the majority of the tooth, and so serves as a grinding edge, particularly where folds are present in the tooth, rather than just at the edges.

The molars of some rodents stop growing and so, if these teeth become badly worn or damaged, the animal will have difficulty in eating. The diet of most rodents, based upon vegetable matter, seeds and, in some cases, even bark, can exert heavy demands on the molar teeth, as well as on the incisors, and may ultimately curtail the animal's life.

Not surprisingly, therefore, in a number of rodents, molars have evolved which continue to grow and so do not become worn down. This arrangement is particularly common in burrowing rodents and beavers, and such teeth have no roots. Dental cement helps to hold these columnar-type molars in place.

JAW ACTION

The presence of powerful, durable teeth is not the only factor enabling rodents to eat a wide range of potentially hard and inaccessible foods. The musculature and structure of the jaws need to be developed so that the teeth can be used to maximum effect. Firstly, the jaws must be arranged in such a way that the incisors can be lifted apart when the molars are in action. Both groups of teeth cannot be used simultaneously. They must not be bound tightly by muscles at the back of the skull.

Traditionally rodents have been divided into three groups, based on the musculature of their jaws, particularly with regard to the masseter muscle. This is the main muscle responsible for pulling the lower jaw up to the upper jaw and also extending it so that the rodent can use its incisors.

There are two branches of the masseter. In the case of the suborder Sciuromorpha (squirrel-like rodents), the deep part of the masseter is responsible for closing the jaws, while the lateral branch, which is attached to the snout in front of the eyes, controls the action of the incisors.

The situation is effectively reversed in members of the suborder Caviomorpha (cavy-like rodents). In this group, the deep branch of the masseter is associated with the gnawing action of the jaws, rather than serving to close them. Its point of insertion is on the snout for this reason.

In the Myomorpha (mouse-like rodents), which is the largest suborder of rodents, both branches of the masseter are attached rostrally, thus reinforcing the gnawing ability of these particular rodents. As a result, rats, for example, are able to gnaw through solid concrete. It is this arrangement of the jaw musculature that has contributed to their success.

The actions of the masseter are reflected in the structure of the skull itself. In sciuromorphs the muscle arrangement is basically similar to

The hare-lip appearance associated with rodents is visible in this Cuban hutia (*Capromys* species). This gives more flexibility when it comes to eating.

that in other mammals, with the muscle attached to the cheek bone. In caviomorphs, this part of the skull is greatly enlarged and there is a large hole present on each side, in front of the orbit, which houses the eye. The masseter runs through this gap from the lower jaw to the sides of the face.

In the skulls of myomorph rodents there is a similar space accommodating part of the masseter, although their cheek bones tend to be smaller than those of the caviomorphs and the space for the musculature is much narrower. In this group the gap, or foramen infraorbitale, is positioned at a more vertical angle, rather than being a continuation of the bone beneath the eye, as in the caviomorphs.

BODY STRUCTURE

The overall skeletal pattern of the world's rodents is remarkably similar and reflects their compact body shape. The greatest variability is apparent at the extremities of the body where the lower limbs and tails may be modified to reflect the rodent's particular life style.

Their spinal column is made up of 12 or 13 dorsal vertebrae, followed by 6 or 7 lumbar vertebrae and 3 or 4 sacral vertebrae that form the tail when this is present. The chest cavity is made up of 9 pairs of ribs, behind which is a further pair of so-called 'floating' ribs that are not bound together.

Most species of rodent have quite short legs coupled with a relatively elongated body shape. There is usually a collar bone or clavicle present at the top of the forelimb, although this is not always well developed. Lower down the limb itself, there is no fusion between the radius and ulna bones, and the elbow joint, formed with the upper bone, or humerus, can rotate freely. This enables arboreal rodents in particular to climb easily.

There are normally five digits on the front feet, although the first digit, corresponding to the thumb, may be greatly reduced in size, or even absent. Many rodents use their front feet to hold their food.

In the case of the hind limbs, both the tibia and fibula are generally not bound together, although they are sometimes joined at their lower ends. The bones of the hind legs can be relatively large in the case of rodents which can leap long distances, such as the jumping mice (Zapodinae).

When walking normally, jumping mice use all their feet but, should they become frightened, they can cover 2 m (6½ ft) or more in a single jump. As well as possessing powerful legs, they have a long tail, which helps them to maintain their balance as they leap. Jumping mice which have lost their tail are often unable to land on their feet or jump as far.

In these rodents the number of toes on the hind feet varies between three and five. Jerboas (Dipodidae) show not only a reduction in the number of these digits, lacking the first and fifth, but their metatarsal bones, forming the feet, are joined together. This adaptation undoubtedly makes it easier for them to hop over the sandy desert areas of Asia, where they live.

The front feet of most rodents are smaller but more agile than their hind feet. The underparts of a brown rat (*Rattus norvegicus*) are shown here.

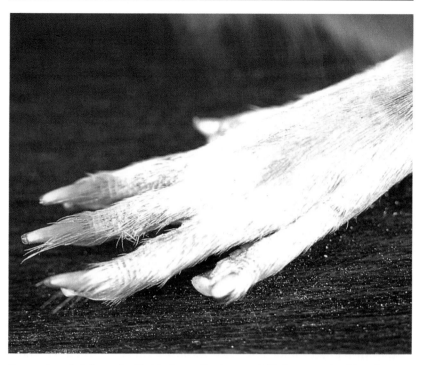

The number of digits on the hind feet vary. The brown rat (*Rattus norvegicus*) has five on the hind feet (above) whereas the guinea pig (*Cavia porcellus*) has just three.

A further adaptation is present on the feet of some jerboas, like the comb-toed jerboa (*Paradipus ctenodactylus*). This rodent has an array of stiff hairs around its toes which serve to increase the surface area of the feet, making it easier to move on slippery sand, when keeping a foothold can be difficult. A similar adaptation can be seen in other jerboas, including the feather-footed jerboa (*Dipus sagitta*).

Those rodents which regularly spend part of their lives in water, like the beavers (*Castor* species), have webbing between their toes, although this is confined to the hind feet in the beavers. They also have sharp claws, with the innermost two on each hind foot being notched to facilitate grooming.

Rodents generally walk on the soles of their feet in what is described as a plantigrade fashion, but those which rely on speed to escape predators, such as agoutis (*Dasyprocta* species), are digitigrade. This literally means that they run on their digits, with this part of the foot supporting their weight. In digitigrades, the claws may help maintain balance when the rodent is running, by sticking into the ground and preventing them slipping.

Dune mole rats (*Bathyergus* species) which live primarily underground, generally have particularly strong claws. These are adapted for digging and are used when constructing subterranean networks of tunnels. In some species, however, like the blind mole rats (*Spalax* species), the front feet are quite small and terminate in short, rounded claws. The prominent incisor teeth are the preferred implement for digging, while the feet act as shovels to scrape the earth away. This is because these rodents live in harder soil than other mole rats.

The body shape of mole rats is also modified to suit their underground life style. Their rounded skull is reinforced and especially strong and they keep their lips closed behind their incisors so that soil does not enter their mouth as they tunnel.

When clearing the dirt from a tunnel, mole rats rely on their front feet primarily for balance, sweeping away the soil with their hind legs and tail. *Tachyoryctes* species use their heads rather like brushes to move the spoil backwards. Backwards kicking is also used to remove earth from the network of tunnels. Both feet work together, creating piles of spoil on the surface. This activity is sometimes referred to as 'volcanoing' and is usually carried out at night, to lessen the risk of the mole rats being disturbed by possible predators. Their compact shape also allows mole rats to move over each other more easily within the tunnels. It is usual for the smaller individual to crouch down, allowing the larger mole rat to move over on top.

In collared lemmings (*Dicrostonyx* species) the claws vary quite significantly through the year in response to environmental changes. Originating in the far north, where thick snow covers the ground in winter, these lemmings have double-pointed claws on the second and third digits of their front feet in winter. This helps them to dig through the snow in search of vegetation to eat. In summer these are replaced by smaller, single-pointed claws.

Tree squirrels (*Sciurus* species) also depend heavily on their claws, which enable them to climb trees without difficulty and descend head-first to the ground. Dormice are also well adapted to a predominantly arboreal existence. The common dormouse (*Muscardinus avellanarius*) can rotate the lower part of its limbs to improve its grip, while its claws are also elongated (with the exception of those on the innermost digit on each foot) for the same reason.

A less obvious modification to the digits is apparent in the viscacha (*Lagostomus maximus*), a South American rodent related to the chinchillas (*Chinchilla* species). It has just three toes on each of its hind feet and the centre toes have a swollen area close to the tip. This is covered with bristles rather than hair and is used for grooming. Tuco-tucos (*Ctenomys* species) have similar bristles, again for grooming, but they are located on the edge of the feet rather than more centrally on the toes themselves.

Certain rodents, notably the spiny rats (*Dactylomys* species), have evolved particularly long toes to assist their climbing activities. In this case there is a space between the elongated third and fourth toes of their front feet which they can use to hold on to a thin branch, anchoring themselves even more securely with the second and third toes of the hind feet.

THE TAIL

In myomorphs, tail length is a very variable characteristic and can be related to the life style of the individual rodent species. The small European harvest mouse (*Mycromys minutus*) uses its tail not just for balancing but also for grasping stalks of corn. The tail therefore serves as an extra limb and provides additional support as the mouse climbs in the grass.

At the other extreme, the Syrian hamster (*Mesocricetus auratus*) has just a basic stub of a tail that appears to serve no functional purpose. Many rats and mice have bare, scaly tails, with a much more limited prehensile function than that of the harvest mouse.

Whereas the beaver uses its tail directly for swimming, other aquatic rodents use their tails as rudders. The tail of the muskrat is flattened on its sides for this purpose.

The presence of a dense covering of fur as a counterbalance, as found in hopping and jumping rodents, like the jirds (*Meriones* species), is a feature developed to an even greater extent in those species which glide from tree to tree. Flying squirrels, such as the southern flying squirrel (*Glaucomys volans*), have a heavily furred tail which serves as a drag anchor as well as providing balance.

Only the cavies (*Cavia* species) have no tail. This is one of the distinctive features of this group of rodents. While most caviomorphs live on the ground, where they may burrow, a number of porcupines, which often climb, notably the South American tree porcupines (*Sphiggurus* species), do possess well-developed tails.

The tail of arboreal rodents, such as this grey squirrel (*Sciurus carolinensis*), is important in providing balance.

In the case of the prehensile-tailed porcupines (*Coendou* species), the tail is equivalent to almost 10 per cent of the animal's body weight and is made up of muscular tissue that enables it to grasp branches effectively. The tip of the tail curves upwards and is free of spines. An area of hard skin gives good contact between the tail and the branch itself.

In conjunction with their tail, these porcupines also have strong feet. The first digit on each foot, although reduced in size, is merged into the pad of the foot itself, which improves the porcupine's ability to maintain its grip. Being relatively large and heavy poses a particular challenge in terms of an arboreal life style.

Perhaps surprisingly, these porcupines rely mostly on their tails and footpads to maintain their balance in the trees, rather than on their sharp claws. They are often to be seen on branches 10 m (33 ft) or more off the ground and a fall from this height could result in serious, if not fatal, injury.

Such is the importance of the tail to arboreal rodents that it is tightly anchored to their body. The African dormouse (*Graphiurus murinus*) typically has a tail which is nearly as long as its body. If it becomes damaged and the tip breaks off, the tail can regenerate, at least partially, rather like that of a lizard. Thus, the dormouse too may be able to escape from predators by shedding part of its tail. A similar phenomenon occurs in other genera of dormice and regeneration of the sacral vertebrae forming the tail, and the intervertebral discs, has been observed.

In certain other species of rodent the skin covering the tail, particularly near the tip, is weak and offers a similar means of defence against predators. The Mongolian gerbils (*Meriones unguiculatus*) is able to shed the skin of the distal portion of its tail with little resulting blood loss.

Studies of the value of this method of protection have been carried out on the Florida mouse (*Podomys floridanus*). Nearly one in ten showed damage to the tail, which had presumably been caused by snakes or other predators.

THE SENSES

The brain of rodents has a relatively simple structure and there are few furrows evident on its surface. Nevertheless, rodents often display a capacity to learn, as shown by the reluctance of rats and mice to take poisoned bait, especially once other members of a colony have died from its lethal effects. In a similar way, rats will learn to avoid traps.

Many rodents have relatively large eyes, a feature common to animals which are most active during the hours of darkness. The relative importance of the different senses, however, depends to a large extent on the rodent's life style, with those which are active during the day relying more heavily on sight than species which live underground. In fact the mole rats (Bathyerginae) have very small eyes, while members of the subfamily Georycinae are entirely blind. In the latter, tiny, non-functional eyes are present but are concealed beneath the skin.

In contrast, rodents living above ground, in open countryside, are particularly vulnerable to birds of prey and rely primarily on keen eye-

Large eyes, like those of this springhare (*Pedetes capensis*), are typical of rodents which are active from dusk until dawn.

sight to alert them to possible danger. Gundis (*Ctenodactylus* species), from arid areas of North Africa, will squeeze under rocks to escape predators. Eyelashes can prevent rodents being dazzled by bright sunlight and maras (*Dolichotis* species), for example, have very long lashes.

Overall, however, hearing and smell are the most important senses in rodents. Gundis rely on their flattened ears to trap any sound waves around them in the still desert air; these can convey the wing beats of an approaching predatory bird or other sources of danger, like snakes. Within the skull itself, the ear capsules are enlarged, as they are in other desert-dwelling species. This serves to reinforce faint sounds. The sensitivity of their hearing is also much more acute than that of human beings. They are able to detect sound frequencies of up to to 40,000 Hz (within the ultra-sound range) that are inaudible to the human ear.

In conjunction with their acute hearing, rodents communicate vocally with each other. Different species of gundis, whose ranges overlap, have different call notes, enabling individuals to recognize members of their own species. They also call to indicate possible danger. Urgent, short vocalizations indicating that a bird of prey has been spotted nearby, will send members of a group scurrying for cover, whereas more protracted calls reveal danger on the ground.

Many rodents, particularly desert-dwelling species, also drum on the ground with their hind feet as a means of communication. In contrast, deer mice (*Peromyscus* species) use their front feet for this purpose.

The sense of smell is important in rodents and their olfactory system is well developed, as can be seen from the large nose of this brush-tailed porcupine (*Atherurus* species).

These are not necessarily alarm calls, but are more likely to be a means of keeping in touch with others of the same species.

Beavers (*Castor* species) hit the surface of the water with their tails as a warning signal. Wood rats (*Neotoma* species) use their long tails like rattles for the same purpose, with the resulting noise being discernible at a distance of 15 m (16½ yd) away.

Although mice and other myomorphs may appear quiet, this is simply because their high-frequency sounds are inaudible to the human ear. If their vocal chords become damaged, however, their calls sound rather like whistles. The accounts for the so-called 'whistling mice' which are reported in the media from time to time.

The noisiest rodents are found in South America. This distribution may have an evolutionary basis, because mammalian predators only invaded this part of the world quite recently on the geological time scale. Clearly, the major risk associated with vocalizations is that the sounds will give away the rodent's position and make it more vulnerable to a predator. (It is no coincidence that servals, for example, inhabiting grassland areas of Africa, have developed a very acute sense of hearing which enables them to locate concealed rodents on the basis of their ultrasonic calls.)

Guinea pigs (*Cavia* species) rank amongst the noisiest of all rodents, often squeaking and squealing simply as a means of communication rather than specifically to indicate possible danger. Their social nature, however, brings extra advantage in terms of more pairs of eyes to detect the approach of potential predators, so that perhaps remaining concealed as much as possible becomes less significant.

BODY COVERINGS

The body coverings of rodents range from bare skin, as in the naked mole rats (*Heterocephalus glaber*), through to the very dense fur of the chinchillas (*Chinchilla* species), in which as many as 80 hairs are produced from a single follicle.

Significantly, naked mole rats still retain their whiskers, or vibrissae, around the face, and these are also present in other rodents. These whiskers have a sensory function, providing the rodent with information, on for example the width of a tunnel, or whether a crevice is large enough to provide a refuge from a predator. They provide close-up sensory feedback, even in the dark.

The viscacha (*Lagostomus maximus*) has long vibrissae, and males, in addition, have a moustache-like array of longer hairs. This is a unique feature among rodents, although why such a means of distinguishing between sexes should have evolved is unclear, as it appears to have no great functional significance.

Many rodents, especially those found outside desert regions of the world, tend to be agouti in colour. This is the term given to the dark and light pattern of banding on individual hairs, which helps to provide a degree of camouflage.

The coat coloration of many rodents helps them to blend in against their background, as shown by this mara (*Dolichotis patagonum*).

Wild cavies (*Cavia tschudii*), for example, are far duller in coloration than their domesticated relatives and the same applies to other domesticated rodents, although some colour mutations have been recorded in wild rodents, notably rats.

Some South American caviomorphs have body markings which give an instant warning to others nearby of any threat. This is most marked in the case of the maras (*Dolichotis* species), which have a white fringe around the base of their rump. This becomes more obvious when these rodents are running.

Desert rodents tend to have lighter-coloured fur than their counterparts elsewhere, which again assists them to blend into their background. The camouflage value of coat coloration is well demonstrated by collared lemmings (*Dicrostonyx* species), which remain active through the cold winter months in the far north. Their coat turns from brown to white in winter, enabling them to blend in against the snowy background. Other rodents, such as the Russian dwarf hamsters (*Phodopus* species), also undergo this change, reverting to their darker coat again in the spring.

The underparts of many rodents are paler than their upper parts, frequently bordering on white. This may reflect heat radiating from the ground and is particularly significant in desert-dwelling species. Such rodents tend to venture out of their burrows after dark, when the air temperature is falling rapidly but the surface of the sand may still be

Desert-dwelling rodents can be recognized by their pale sandy coats, which help them to merge into the sandy landscape. This is a fat sand rat (*Psammomys obesus*), which originates from North Africa and the adjoining area of the Middle East.

hot. Some rodents living in deserts may also cool their bodies by smearing saliva over their faces with their paws.

Apart from the adaptive coloration of their fur, desert rodents may lack fur on their extremities of their bodies, such as their tails and ears. Their ears are also frequently enlarged so that heat can be rapidly dissipated by an increased blood flow.

The body covering plays a vital part in assisting flying squirrels, like *Petaurista* species, to glide. For this purpose, there is a fold of skin running along the sides of the body between the front and back legs, or even between the neck and tail, and attached to the wrist by a bony spur. When the squirrel is climbing normally, this membrane is inconspicuous but, when the wrists are extended, it becomes visible.

The squirrel is able to control its movements while in flight by adjusting the amount of contraction within the sheets of muscle comprising the membrane. It relies on its tail as a rudder and for braking its glide. As the squirrel lands, it raises its tail, increasing the drag factor and enabling it to land with its claws in a position to grasp the branch or trunk.

The power of these membranes, and their structure, are not sufficient to allow this group of squirrels to fly in the true sense, however; inevitably, they glide downwards rather than to a higher landing point. Even so, there is a safety factor built into their gliding skills: the extended membrane can also act like a parachute, slowing their descent to a safe level. Flying squirrels have been known to drop at least 180 m (590 ft) and avoid any obvious injury.

Undoubtedly the most distinctive group of rodents, by virtue of their body covering, are the various porcupines. Variable numbers of their hairs have become modified into sharp spines, although they still possess an undercoat of hair and some longer hairs.

Up to seven different types of bristles and spines have been identified, of which the most dangerous are the spikes. These are also the thickest and can reach a length of 25 cm (10 in). Much of the porcupine's body may be covered in sharp spines, which are flattened in shape and also quite stiff.

More localized are the tactile bristles, which have a sensory as well as a defensive function. These may be longer than the spikes, reaching a length of 45 cm (18 in). Although they are far more flexible, they still have sharp tips.

The porcupine's neck and head may be covered with bristles which can grow to 7.5 cm (3 in) long, while the tail is equally well protected by a range of bristles. There may also be so-called rattling cups attached to the tip of the tail, which are used by some species to make a distracting noise, aimed at deterring potential predators.

The degree of protective covers differs markedly among the porcupines. The long-tailed porcupine (*Trichys fasciculata*), found in parts of South-East Asia, carries a very sparse armoury of spines and bristles. Its tail is scaly, with bristles being confined to the tip. The tail itself is very fragile and the tip breaks off easily. The head and underparts show no

The hair of a number of rodents has become modified to form bristles, spines and quills, as in this African porcupine (*Hystrix cristata*).

The tail may offer protection, as in the case of this brush-tailed porcupine (*Atherurus* species). The bristly tip of the tail breaks off easily, allowing the porcupine to escape if caught by its tail.

signs of bristles, being covered with hair. Flattened spines protect the rest of the body.

The more highly evolved brush-tailed porcupines (*Atherurus* species) show greater protection along their back, with their whole coat covering being spiny.

Rattle quills are present in the porcupines of the genus *Hystrix*, which includes the most formidably armed members of the family. African species, such as the North African crested porcupine (*H. cristata*), have a mane of long, thick bristles on the head and neck, and particularly elongated, sharp bristles covering their hindquarters and extending down the sides of the body. These porcupines raise their spines if threatened, which makes them an imposing sight, and rattle their tails. Grunting and snarling follow and the porcupine also drums on the ground with its hind feet.

Should this prove to be an insufficient deterrent, the porcupine will attack, launching at its target, either backwards or sideways, to make the most painful impact with its spines. The spines break off after becoming impaled in the target's body, enabling the porcupine to retreat to safety as its potential attacker is painfully distracted. It is necessary for the porcupine to make contact with its attacker, however; contrary to popular belief, the spines cannot be projected out of its skin.

The length of the porcupine's quills is not necessarily a reliable indicator of their effects. The North American porcupine (*Erethizon dorsatum*) may have more than 30,000 quills protecting its body, each of which is up to 7.5 cm (3 in) long. In true porcupine fashion, it defends itself by lashing out with its tail, leaving behind a number of these quills, which are just 2 mm (⅟₁₆ in) in diameter and have barbed tips. The skin contractions at the site of the wounds draw these tiny darts deep into the dermis, at a rate of about 1 mm (less than ⅟₁₆ in) per hour. It is not unknown for these quills to penetrate vital blood vessels or body organs, proving fatal within hours, if not days, of the original encounter.

THE DIGESTIVE SYSTEM

Although the external appearance of rodents may appear to be generally similar, the structure of the digestive tract can differ quite widely in different species. Rodents, like rabbits and other herbivores, face the problem of digesting plant matter. The cell walls of plants contain cellulose and no mammal has evolved the ability to digest this constituent by means of its own enzymes. Instead, mammals are forced to rely on helpful microbes located within their digestive tract, which possess cellulases for this purpose. Unfortunately, these bacteria and protozoa are located in the caecum, at the junction of the small and large intestines, which is beyond the point at which foodstuffs can be absorbed into the body. The practice of refection, therefore, has evolved.

Foodstuffs pass through the stomach and small intestine as far as the caecum, where the bacteria break down the cellulose. Soft faecal pellets are then passed out of the anus. These are consumed by the rodent and

Some rodents, such as this agouti (*Dasyprocta* species), will feed during the day but the majority prefer to forage for food after dark.

If caught in the open, rodents will immediately seek cover, down its burrow in the case of this springhare (*Pedetes capensis*).

the pre-digested cellulose, in the form of its carbohydrate constituents, is absorbed into the body through the walls of the small intestine. The water content is then absorbed in the large intestine and the rodent then passes hard, faecal pellets. Foodstuffs which do not contain cellulose can be absorbed without passing through the intestinal tract a second time.

Feeding can be a very dangerous time for rodents because they are usually exposed, in the open, and are potentially vulnerable to predators. As a result, a number of rodents have developed cheek pouches. These enable them to gather food in quantity without pausing to eat it above ground. The common hamster (*Cricetus cricetus*) has been known to amass as much as 90 kg (198 lb) of food in its burrows, all carried to this store in its pouches.

Both pocket mice (*Heteromys* species) and pocket gophers (Geomyidae) have pouches similar to those of the hamster. These are made up of folds of skin lined with fur and may extend back as far as the shoulder region. They are controlled by a special muscle that enables the pouch to be everted as required.

SOCIABILITY

Where conditions are suitable, some rodents live in colonies. This sociability has benefits both for the individual and the group, and helps to contribute to the survival of the species. A single beaver, for example, would have more difficulty in constructing a dam and lodge than a pair or a larger number cooperating for this purpose.

This learning process is also assisted by association, as mentioned previously in the case of rats and mice, where bait shyness can develop throughout a colony (see page 38). The likelihood of reproductive success is also maximized, since rodents can breed rapidly under favourable conditions.

Those rodents which live on their own are often species occurring in desert areas, where food resources tend to be most limited. Under these conditions a lower density population, spread out over a wider area, clearly brings survival advantages.

Rodents themselves are highly significant in the ecosystem of the desert, but not only as prey for other animals. Their tendency to burrow affects the vegetation as well as the soil. Wherever rodents are well established, plant cover tends to be more evident. This is partly because seeds are carried into the area and partly because the soil becomes dug into mounds and it is easier for seeds to germinate.

Unfortunately the effects are not entirely benign since rodents' burrows may be invaded by a host of other creatures seeking sanctuary. Some, such as arthropods, including fleas, represent a potentially serious disease reservoir. Other vertebrates, including green toads (*Bufo viridis*) and various reptiles, as well as small insectivorous mammals, like various shrews and hedgehogs, may share the rodents' burrows.

Typical of these desert-dwelling rodents is the great gerbil (*Rhombomys*

The way in which rodents associate can be influenced by their environment. Those which live in areas where food resources are scarce, such as desert regions, tend to be less social than those from areas of forest, such as the brush-tailed porcupines (*Atherurus* species).

opimus), which occurs over a wide area of central Asia eastwards to northwestern China and Mongolia. Where burrowing conditions are favourable, a network of burrows, with a number of different entrances and galleries connecting the chambers, can extend over an area of 1 km² (about ½ sq. mile). As many as 500 other species of both invertebrates and vertebrates may be found in these burrows, which demonstrates the importance of these gerbils within the ecosystem of these regions.

The ability of the great gerbil and similar rodents to survive under harsh desert conditions is enhanced by a range of physiological adaptations. Studies have revealed that their metabolism slows down when conditions are unfavourable, reducing their energy requirements. Their oxygen requirement, for example, is lower than that of rats and they are better adapted to survive periods of dehydration.

Water conservation is clearly a priority for desert-dwelling rodents and their kidneys are incredibly efficient at conserving water. This is thanks to the very long loop of Henlé in each kidney. This structure enables a large volume of water to be reabsorbed into the body as part of the filtration process, resulting in the production of a very concentrated urine.

Although there may be little free-standing water in desert areas and rainfall is sparse and erratic, condensation forms in the entrances to the burrows in the early morning, providing a meagre but dependable source of fluid for the rodent inhabitants. A few rodents, such as the fat sand rat (*Psammomys obesus*), never actually drink, but meet their fluid requirements from plants and other food instead.

Rodents living in Arctic areas face an equally inhospitable landscape and a different set of survival problems. There, the lower layers of the soil are likely to be permanently frozen throughout the year and this permafrost makes burrowing in order to survive a difficult option.

Only the top part of the soil thaws out, even in summer, and this thaw is likely to result in flooding because water cannot drain into the permafrost layer beneath. During the winter the soil is frozen solid and a layer of snow accumulates on top. It is actually the snow, which is soft and provides insulation, that enables rodents to live in the Arctic throughout the year.

The frozen ground enables rodents to extend their feeding areas into sites which would be too marshy for them in the summer. Lemmings (*Lemmus* species), which frequent this part of the world, actively seek out the areas of deepest snow in the winter, feeding on the buried vegetation. Here, and at this time, just under half the female lemmings will breed, which demonstrates the protective effects of the snow. Breeding ceases in the spring, as the snow melts and areas become flooded, and starts again during the summer months, when it reaches a peak.

The fat sand rat (*Psammomys obesus*) lives in North Africa, including parts of the Sahara desert. It is a desert-dwelling rodent that may be forced to consume succulent plants with a high salt content. This vegetation provides a supply of water, with the kidneys excreting the excess salt from the body in a very concentrated urine which conserves body fluids.

A number of rodents found outside the tropics, such as the grey squirrel (*Sciurus carolinensis*), lay down winter stores of nuts, which will sustain them when other foods are in short supply. Some caches are invariably lost or forgotten and this enables the nuts to germinate in new areas, thus helping trees to spread.

A different life style is adopted by the Arctic ground squirrel (*Spermophilus undulatus*). These rodents can only be found in areas which do not flood and they instinctively favour higher stretches of ground where they can excavate permanent burrows, which they fierce-ly defend. Ground squirrels have also adapted to the terrain in terms of diet and eat a much wider range of foods than their counterparts in more southerly latitudes. In addition to grasses and other plants, Arctic ground squirrels regularly eat insects.

In temperate parts of the world, rodents have their greatest impact in grassland areas. Not surprisingly, those which spend almost their entire lives below the soil surface, such as the blind mole rats (*Spalax* species), are likely to have a significant environmental impact. These mole rats are each capable of shifting as much as 15 m³ (20 cu. yd) of earth per hectare (2½ acres) of ground annually.

Brought to the surface and piled up, this earth is significant in terms of the moisture retention of the ground. Studies have shown that the effects can be apparent for at least 3 weeks after a period of rainfall. The mounds of earth influence the run-off of rain and are also more absorbent than the ground. Thus the soil immediately around the mound becomes damper than that further away. The burrowing activ-ities of rodents can also bring minerals closer to the surface, improving the fertility of the soil. They can also shift humus, which itself has water-retaining properties, deeper into the earth.

The effects of rodents in areas currently under cultivation, however, are reduced because there is little opportunity, especially under intensive agricultural conditions, for colonies to become established. Also, the pattern of land management, often coupled with rodent control measures, means that the population will remain low.

Rodents have less direct effects in areas of mature woodland, in terms of soil quality, than in grassland. Nevertheless they can still affect the landscape by feeding on seeds, often carrying them further afield where some may be left to germinate. Overall, the destructive effects of rodents in woodland are generally less than might be expected, although, in localized areas, they may result in heavy financial losses.

Newly planted areas of woodland are the most vulnerable to the deprivations of rodents because, in addition to eating seeds, rodents are liable to consume seedlings. These have a higher nutritional content than older trees. Tree seedlings which do survive the attack of rodents are ultimately less likely to attain their full height on maturity.

The greatest diversity of rodents is found in tropical regions, where their populations appear to be more stable than in other parts of the world. This is probably because of the relative stability of the climate, although population peaks have been recorded in some species.

Huge localized increases in the numbers of rodents may be related to changes in the vegetation, often as a result of agricultural development.

Some rodents have a restricted diet and this influences their distribution. The common dormouse (*Muscardinus avellanarius*) is found in areas of woodland where hazel grows. These nuts form an important part of its diet in the autumn and the characteristic teeth marks on the shells enable estimates of dormice populations to be made.

The grazing habits of some rodents such as the capybara (*Hydrochaeris hydrochaeris*), can alter the pattern of vegetation in an area, especially when their numbers are high.

The wood rat (*Rattus tiomanicus*) has built up from low numbers to densities as high as 500 per hectare (2½ acres) in oil palm plantations, where it feeds on the nuts and causes serious economic damage. It has been necessary to encourage barn owls (*Tyto albo*) into these plantations, by providing artificial roosting and nesting sites, to curb the wood rats.

Breeding of rodents in the tropics is variable. Some breed throughout the year, whereas the breeding period for others is closely related to the rainy period. Again, however, the greatest number of species is found in grassland areas rather than in tropical forests.

THE DRAWBACKS OF BEING A RODENT

While the small size of rodents helps them to avoid potential predators and to conceal themselves easily, this is achieved at a price. Because their surface area is large in relation to their volume, rodents lose heat more rapidly from their bodies than larger mammals. To maintain their body temperature, their metabolic rate has to be increased substantially. Fuelling this increase places large demands on their food-gathering abilities and also increases their demands for oxygen. It has been calculated that some of the smallest rodents require an increase of up to

100-fold in both nutrients and oxygen, compared with large mammals whose metabolism can function at a more sedentary rate.

Rodents may be forced to find and eat approximately 70 per cent of their body weight every day. When conditions are unfavourable, the metabolism of some species, such as the Syrian hamster (*Mesocricetus auratus*), may slow down so much that they become torpid. A decrease in temperature is known to induce this torpor in Syrian hamsters, which have been recorded as surviving 28 days in this state, sustaining themselves on body-fat stores.

In temperate regions, some rodents, like marmots (*Marmota* species), spend much of the winter hibernating underground in their burrows, rather than battling with the elements to find food. In summer, they start to build up reserves of fat which, by autumn, may account for 20 per cent of their total body weight. Retreating underground, marmots may spend as long as 9 months in hibernation. They occasionally emerge from their burrows for a brief period during warm spells. In the southern part of their range, in mild winters, woodchucks (*M. monax*) may either not hibernate at all or spend just a few weeks below ground. This is yet another example of the adaptability of rodents.

Chapter 3
Reproduction

Their reproductive potential is one of the reasons why rodents have proved to be such a successful group. Many species, particularly among the myomorphs, have a short generation time and large litters. This enables them to adapt rapidly in the face of environmental change, whether favourable or potentially damaging. The brown rat (*Rattus norvegicus*), for example, has spread rapidly into new localities, as well as acquiring resistance to certain rodenticides within a relatively short space of time.

Against this background, it is easy to overlook the fact that not all rodents are prolific breeders. The mountain beaver (*Aplodontia rufa*) has a short breeding season in spring, after which the female may give birth to just a couple of offspring.

The offspring are also slow to develop in rodent terms, remaining in the nest until they are 2 months old. Subsequently, they will probably not breed until they are 3 years old. By way of compensation, the life expectancy of the mountain beaver is far greater than that of many other rodents, being about 6 years according to some estimates.

The way in which rodents breed can be influenced by their current population and whether or not they are social by nature. This group of spiny mice (*Acromys* species), like most myomorphs, can produce large numbers of offspring in a relatively short space of time.

Most rodents give birth in a nest or underground burrow. This agouti (*Dasyprocta* species) is about to enter its burrow concealed at the base of these rocks.

This primitive rodent is in decline, however, being the only surviving member of a much larger group, known from the fossil record, whose remains have been found in both Asia and Europe, as well as North America where it occurs today. This suggests that the capacity to increase their reproductive potential has been a major factor in the success of rodents, notably within the prolific myomorph group.

DISPLAY AND COURTSHIP

Overt signs of courtship are most apparent in diurnal rodents, such as squirrels. Here the male will pursue his intended mate, pausing on occasions and using his bushy tail to signal his intentions to the female. He flicks it in an upward direction over the head. As the male continues his pursuit he makes calls similar to those produced by young squirrels, which appeal to the maternal instincts of the female. In due course, she slows down and ultimately the pair mate.

Apart from visual signals, odours also play a very significant role in the reproductive behaviour of rodents. In the case of social rodents, such as mice, there is still a distinct order of dominance with a colony. Males, typically the largest and strongest ones, are dominant.

On occasions, advertising may be necessary to attract potential mates. Male mice use their pungent urine for this purpose. Recent studies have shown that the urine also contains an ultraviolet component, which may help females to locate the male in a particular area. Of course, it may also allow predators to track them but probably no more effectively than by following the scent trail.

Rodents which are naturally antagonistic towards each other, like the Syrian hamster (*Mesocricetus auratus*), must also indicate their presence to a suitable suitor when they are ready to mate. In this species, the female also uses scent-marking for this purpose, relying on her vaginal secretions rather than urine. She produces a specific odour approximately 2 days before she is ready to mate, increasing the likelihood of attracting a male at the appropriate time. Should her quest prove unsuccessful, however, the female hamster then ceases this behaviour until she again reaches this stage in her reproductive cycle.

The urine of male mice actually serves to bring mature females into breeding condition and can speed up the onset of maturity in younger individuals. By repeated scent-marking in an area where there are few mice, a male can start to exert an immediate effect on the population. The likely result will be a rapid increase in numbers, in a short space of time, allowing the rodents to exploit favourable environmental conditions to maximum effect.

Experiments have shown that a neutered male has no effect on female mice, so it is probably the presence of the hormone testosterone in the urine which is responsible for these effects. However, this situation does not apply to all myomorphs; for example, no similar effect has been noted in rats.

Other changes are likely to follow the introduction of a new dominant male into a group of mice. Females which have recently become pregnant will reabsorb their embryonic offspring. They will then come into heat almost immediately and mate with the newcomer.

This biological response may seem potentially damaging to the well-being of the colony but, in fact, it could well prove to be beneficial. By mating with as many females as possible almost immediately after joining the group, the male's genes will be widely distributed. Even if he dies soon afterwards, he will leave descendants.

The survival value for the group lies in the fact that this response by the female mice helps to reduce the risk of in-breeding in a population. Since one of the most likely effects of matings between closely related individuals over a number of generations is a reduction in litter sizes, this could ultimately leave an entire population at risk.

The impact of in-breeding would quickly become apparent in prolific rodents such as mice, which are capable of producing a large number of offspring in a short period of time. Every opportunity to expand the gene pool is therefore potentially valuable.

There is a sophisticated feed-back system which may well start with the female mice in a group. Their urine contains a chemical messenger, or pheromone, which acts on the male, stimulating the production of his pheromone. This in turn triggers female reproductive activity.

When female mice are on their own, however, their pheromone appears to have the effect of prolonging their oestrous cycle beyond the normal 4 days. The addition of a male will almost immediately return this cycle to normal, by means of the pheromone present in his urine. The females' cycles also become synchronized, ensuring that they are ready to mate at the same time.

Again, this is highly advantageous in the case of a species whose members have a short life span and face many potential predators. The male house mouse (*Mus musculus*) is able to mate successfully with as many as 20 females in a period of just 6 hours, so that no time is wasted. The frequency of mating while the female rodent is in oestrous is quite remarkable. In fact, Shaw's jird (*Meriones shawi*) can mate over 200 times in a day, which is more than any other mammal.

Although much still remains to be learnt about the individual reproductive triggers for many rodents, it is clear that some possess a further means of ensuring maximum reproductive success in the shortest possible time. This is achieved by the female undergoing what is known as a post-partum period of oestrus almost immediately after giving birth.

By mating again at this stage, the female will be carrying her second litter while also caring for her earlier offspring. If the young rodents are taken by a predator or otherwise die during this period, then the next litter will be born after a short interval. This may allow the female to take advantage of favourable conditions for rearing her offspring without the need for her to mate again.

Should the first youngsters continue to thrive, and escape the attentions of a predator, then the development of the next litter will continue to be delayed hormonally until they are close to weaning. The female's second pregnancy will then proceed as normal. Again, it tends to be the more prolific rodents in which post-partum oestrus has evolved.

As the population of rodents in an area grows, obvious behavioural changes may well be observed. Male house mice (*Mus musculus*) living at low population densities tend to be far more territorial and aggressive than those living in crowded surroundings. This is when an order of dominance is established, with fighting being avoided by a range of coded appeasement gestures. They are made by subordinate members of the colony when in a potential conflict situation, or when challenged by another group member.

The need for this shift in behaviour is probably linked to the reproductive capacity of the species. An area can rapidly become over-run with these rodents and aggression under these circumstances would be counterproductive. House mice populations have been known to rise on occasions up to nearly 200,000 individuals per hectare (2½ acres).

However, in contrast to the recognized cycles observed in species such as various lemmings (see page 79), this is not a regular cyclical event but simply a response to favourable environmental conditions. Even so, the effects on a region suffering a plague of rodents can be devastating, as they will consume almost anything edible. Farm crops are often a contributory factor to such population explosions because they represent food supplies at a much greater density than would be available under natural conditions.

Newborn rodents will seek out the warmth of their mother's body. They are at even greater risk of hypothermia than adult rodents. While she is suckling her litter, a female may already be pregnant again. A post-natal period of oestrus is common in many species, including Shaw's jird (*Meriones shawi*), shown here.

The reproductive capacity of rodents relies not only on the frequent production of large litters but also on the early maturation of the young. The meadow vole (*Microtus agrestis*) matures at just 25 days old and ranks among the most prolific of all rodents. Females can subsequently produce as many as 17 litters, each comprising up to eight young, in the space of a year.

Pregnancy can, however, prove to be a dangerous period for female rodents. Their already high nutritional requirements are made even greater by the need to nourish the developing embryos. This means foraging for longer periods in search of food, increasing the likelihood of the mother falling victim to predators.

Although the major growth period of the embryos does not occur until the last part of pregnancy, their weight can prove to be a fatal restriction on the female's movements if she is being pursued by a predator. Thus, although the reproductive potential of rodents is generally high, so is the level of mortality, beginning before the young themselves are born.

BIRTH AND DEVELOPMENT OF THE YOUNG

Many rodents have a short gestation period; that of the Syrian hamster (*Mesocricetus auratus*), for example, typically lasts about 16 days. The young are helpless at birth, being naked and blind. Their eyes start to open about 12 days after birth, by which time the young hamsters have a good covering of fur. They emerge from their nest for the first time soon afterwards and are fully weaned by the age of 1 month.

Scent remains very important throughout the period that the young are in the nest. By transferring her scent to her offspring, the female rodent reduces the likelihood of them being attacked by a male. This is because the urine of female rodents appears to contain a substance which results in a decreased level of aggression among males, who, as a consequence, rarely attack females.

Glands on the rodent's body have a similar function. In the Syrian hamster, it is the lateral scent glands that serve this purpose. The development of these glands appears to be related to the hormonal output of the ovaries.

In the Mongolian gerbil (*Meriones unguiculatus*), the mid-ventral glands appear to serve the same purpose. Again, they only develop after the onset of sexual maturity, which means that, until this time, young gerbils are unable to produce a strong scent and so are not seen as a direct threat to mature males.

A means of reducing aggressive encounters is perhaps more significant in the case of more solitary rodents. Although female Chinese hamsters (*Cricetulus griseus*) will attack males, by biting at their genital region, this behaviour usually ceases if the male rolls over onto its back, revealing the prominant glandular area on its underside. Sniffing at this area appears to placate the female.

Even so, there may be occasions when litters are cannibalized. This often appears to be the result of repeated disturbance, or may sometimes be linked with a shortage of food or water. In fact, rodents inhabiting arid areas may face particular problems when they are lactating, since this increases their fluid requirement considerably.

It is probably partly for this reason that some rodents have evolved a different reproductive strategy: the production of precocial young. Although the gestation period itself is lengthened, to 150 days in the case of the capybara (*Hydrochaeris hydrochaeris*), this is counterbalanced by the fact that the young are born fully developed.

Young capybaras are miniatures of their parents, able to follow their mother and graze grass almost as soon as they are born. Nevertheless, she may allow them to suckle for approximately 4 months. She adopts a sitting posture for this purpose, out in the open. This enables her to remain alert to the approach of predators while the advanced state of development of her youngsters means that they are more capable of escaping if danger threatens. The fact that young capybaras can feed themselves virtually from birth means that they have a chance of survival should their mother be killed. In contrast, young newborn mice, for example, will inevitably die under such circumstances.

Many of the rodents found in South America produce precocial offspring. This can be linked to the fact that they tend not to burrow nor to adopt tree hollows as nesting sites and it is in the caviomorph group that precocial young are common.

Rodents differ significantly in the degree of their development at the time of birth. While caviomorphs are generally born in an advanced state of development, myomorphs, such as this Shaw's jird (*Meriones shawi*), from North Africa, has young that are naked and helpless at birth.

But there is an important distinction between rodents and the large herbivores that also produce precocial young. The young rodents do not immediately follow their mother but remain quiet and wait for her to find them. They do show some instinctive behaviour, however, as in the North American porcupine (*Erethizon dorsatum*), whose young can instinctively lift their quills within hours of birth. At this stage these quills are soft and shorter than those of adults. As with other rodents, the female porcupine keeps in touch with her offspring by means of special contact calls.

The suckling interval of caviomorph rodents varies quite widely from about 2 minutes to 35 minutes or more, although the youngsters do not suckle continuously. It appears that species which produce their offspring in underground burrows, such as the degus (*Octodon* species), invariably spend longer suckling them than, for example, the caviids, which remain vulnerable in the open.

Degus may also evolve a teat order which assists rapid suckling. In the acouchis (*Myoprocta* species), for example, which occur in the northern part of South America, the youngsters can recognize where they should suckle. In these species, as in other caviomorphs, suckling is only necessary for 2 weeks, although it invariably extends beyond this period.

This helps to maintain a link between the mother and her offspring and may serve a protective function, as she can protect them from danger if they remain in her vicinity.

The females of a number of species are known to be ill-disposed towards their partners after giving birth, although the males are unlikely to harm their offspring. This generally occurs in rodents which do not normally tolerate any close contact. In certain instances, male rodents may even defend their young. Such behaviour has been observed in Cuban hutias (*Capromys pilorides*).

Although it is generally true that caviomorph rodents are slow to mature sexually, there are some exceptions, perhaps most notably the cuis or yellow-toothed cavies (*Galea* species). In these, the average age of maturity in females is about 2 months but, exceptionally, they have been known to become pregnant when just 9 days old. No other mammal matures earlier. The onset of maturity is induced in this case by the presence of a male.

In female rodents, as in other mammals, there is a membrane present in the vaginal opening. One of the features of rodents, however, is that,

A young mara (*Dolichotis patagonum*) with its watchful mother. These rodents are born in an advanced state of development, being living miniatures of the adults. The average litter size is correspondingly low, comprising two youngsters on average.

rather than being destroyed at the first mating, this membrane closes again at intervals for the remainder of the rodent's life. In rodents, it dissolves at the start of oestrus and then reforms during anoestrus. This sealing of the vagina has been noted in all three rodent groups. (One exception is the coypu (*Myocastor coypus*) in which no membrane of this type is found once the first mating has taken place.)

As the time for mating approaches, the membrane breaks down, but the length of time for which this opening remains varies, according to both species and the individual. The position and structure of the membrane also differ according to the rodent concerned. For example, it is much more well developed in the chinchillas (*Chinchilla* species) than in the degus (*Octodon* species).

The vulva itself changes little when the female rodent is in oestrus, although some degree of swelling may be observed in a few species, like the casiragu (*Proechimys* species) and plains viscacha (*Lagostomus maximus*).

In the case of male rodents there is often a bone, or baculum, in the penis and the testes are retained within the lower abdomen rather than being kept externally in a true scrotum. In male rodents, the spacing between the urinary and anal openings is generally wider than in the female of a species, although there is some intraspecific variation.

The testes atrophy in both myomorph and sciuromorph rodents, which are seasonal breeders, but no change of this type has been observed in members of the caviomorph group, even though the female may only be reproductively active for part of the year. Some rodents, such as chinchillas (*Chinchilla* species) have relatively large testes compared with others of similar size, although the reason for this is unclear. It could be linked in some way to the high altitude at which these particular rodents live.

POPULATION DYNAMICS

The reproductive rate in a rodent population is likely to have a direct effect on the numbers of rodents in an area. This may often be closely related to the availability of food, as has been shown by a detailed study of wood mice (*Apodemus sylvaticus*) carried out over a period of nearly 20 years in woodland forming part of the University of Oxford's Wytham Estate in the UK.

This revealed that the single most important factor determining the levels of the wood mouse population in the area was the availability of food, particularly acorns, which serve as the main source of food for these mice over the winter period.

The effects of the acorn crop on the numbers of mice continued through the spring period. Perhaps surprisingly, this was the stage at which the population, having remained quite stable over the winter period, began to decline. If the crop had been poor during the previous year, the numbers of mice fell more sharply in the spring.

Although the wood mice began to breed again in April, relatively few of these early youngsters appeared to survive, or they were driven out by older, dominant males. The population build-up which followed later in the year could be directly related to the number of mice present in the summer. If this figure was high, then the population grew more slowly through to the winter, whereas a low summer figure meant that wood mice numbers started to climb earlier in the autumn.

It is unclear why there should be such a sudden increase in the population during this part of the year. Researchers suggested that perhaps the older males increasingly died off at this stage, and that this would be of more significance if the population was relatively low.

There may also be a dietary link because, although plants start to grow rapidly in the spring, they do not start to fruit until significantly later in the year. Therefore, both pregnant mice and their weaned offspring would probably find it harder to locate food during spring and early summer than in autumn.

Irrespective of when the population build-up begins, however, it is interesting to learn that the number of wood mice found overwintering remained remarkably constant throughout the study. This was in spite of the fact that these rodents could be expected to be more vulnerable to potential predators when there was less cover available over the winter period.

Another study in the same area, over roughly the same time span, compared the numbers of wood mice with those of the bank vole (*Arvicola terrestris*). There were consistently more voles in the woods and particularly high figures were recorded at intervals of approximately 2 to 4 years until near the end of the survey.

The close relationship between the population and the availability of food was emphasized by the fact that, in the spring following a particularly large crop of tree seeds, the numbers in the rodent population were significantly higher than expected. This glut of food had enabled the breeding season to be extended into the winter period.

More definite population cycles have been observed in other rodents, notably the microtine voles and lemmings, especially in more northerly latitudes. The Scandinavian populations of the field vole (*Microtus agrestis*), for example, are known to show cycles which peak every 3 or 4 years. Their numbers can range from just 5 to up to 60 per hectare (2½ acres) over this period.

In the case of the field vole, both mature males and females have fixed home ranges but only males display a strong territorial instinct. Females, at least in some other microtine species, display dominant tendencies. No strong pair bond is formed and, as the numbers of these voles in an area increase, so outbreaks of aggression become more common, with males often bearing the scars of such encounters.

This aggression probably serves to disperse the population, with juvenile field voles being nomadic by nature. This helps to safeguard food supplies in an area, because the population fans out, preventing depletion of food stocks. As a consequence, the period of decline may be less sudden and spectacular than in the case of lemmings (*Lemnus* species).

A study of the meadow vole (*Microtus pennsylvanicus*), which is widely distributed from Alaska to northern parts of Mexico, showed that between 50 and 70 per cent of the population dispersed when the density was high but the figure fell back sharply to between 12 and 15 per cent when there were relatively few meadow voles in a region.

Declines may be related specifically to a lack of vegetative cover as well. In parts of the western USA, the number of meadow voles falls dramatically in regions where livestock grazing keeps the grass short. As a consequence, voles may be driven out from habitat that is well suited to their requirements into areas which are less than suitable. This results in an increased level of predation and thus an overall decrease in the population.

Since it is the younger animals which tend to be driven out, so it is likely that the reproductive capacity of the population will decline, as the established voles become older and their offspring suffer greater predation. Furthermore, there is clear evidence that, although predators exert less influence on the numbers of the vole population as it approaches its peak, the percentage caught will increase subsequently as the population begins to fall. It may be that predators become more attuned to hunting voles and therefore do so with greater success.

Lemmings (*Lemnus* species) differ significantly from voles in at least one major respect: they have two distinct reproductive periods each year, producing young in both the summer and the winter months. Between these times, the terrain is unsuitable for this purpose: the ground is likely to be flooded in spring and freezing, yet without snow cover, in autumn.

Like many other rodents, lemmings have a short life expectancy. At the start of the summer the majority are adult, meaning over 2 months old. Subsequently, at the end of the period leading into winter, a large number of young lemmings has been produced and the percentage of adults in the population will have fallen.

Lemmings display a tremendous reproductive potential, with females sometimes conceiving when just 2 weeks old. They can bear as many as 13 offspring in a single litter and, in a period of 6 months, a pair may produce eight litters – equivalent to a maximum of 104 youngsters – in succession.

As the population expands, so the demands on the meagre food resources become greater until there is simply insufficient food available. This causes hoards of lemmings to leave their normal range and invade neighbouring areas in search of more food. The majority are likely to die.

These large movements of lemmings in parts of northern Europe have captured the imagination of people over the course of centuries, particularly the way in which lemmings may tumble over cliffs in frantic droves and end up being swept out to sea and drowned.

It seems that at least part of this bizarre behaviour occurs because the lemmings have eaten toxic plants, after exhausting their regular food supply. The first known portrayal of the mass migration of Norwegian lemmings (*L. lemmus*) dates back to 1555. People believed that the huge hoards formed in the clouds and fell like rain to the earth during storms.

Overcrowding is also an important factor in these movements and may, in some instances, be more significant than a shortage of food. This is borne out by the predominance of young lemmings in the hoards which fan out from their traditional homeland.

Lemmings from different areas may be forced together by the topography of the landscape, following the easiest route until they find themselves trapped, usually by water. Under normal circumstances, lemmings can swim quite well but, in their drug-induced state, and with mass hysteria pervading their numbers, they plunge into the water from which many are likely to be unable to escape. Thousands of corpses may then accumulate in some areas.

It is an interesting fact that peak years for lemmings are also mirrored by a huge population growth in other species which live in these areas. Voles, butterfly caterpillars and even capercaillies seem to thrive in such years and this is perhaps a reflection of more favourable climatic conditions in the preceding months. Should global warming prove to be a long-term phenomenon, then the likelihood is that the interval between these population peaks will inevitably decline.

Predators also benefit from an increase in the number of these rodents and the population cycle of the Arctic fox (*Alopex lagopus*) is closely tied to that of the brown lemmings in this region. Here, a 30-year study has revealed that both species show a 4-year cycle, with the fox population peaking at the same time as that of the lemmings, or just afterwards. The litter size of these foxes plummets after a crash in the numbers of lemmings.

In the parts of its range where it hunts lemmings, white-coated individuals are predominant in the Arctic fox population. This may make it easier for the foxes to catch these rodents. Elsewhere the blue variety of fox is more common, possibly because coat coloration is less significant when it comes to scavenging for food.

It is not just the Arctic fox which depends on lemmings as a major item in its diet. Populations of snowy owl (*Nyctea scandiaca*) are also affected when lemming numbers crash. This has been shown by studying the records of North American taxidermists. During the winter of 1926/7, for example, large numbers of these owls were forced to move southwards in search of other prey after a collapse in the lemming population. Today, the relics of this mass exodus can be seen in many museums. It is estimated that up to 5,000 snowy owls were shot as they moved to more densely populated areas of North America in their quest to avoid starvation.

RODENT LIFE SPANS

The life span of most rodents is likely to be measured in months, rather than years, although there are exceptions. The oldest recorded rodent was a Sumatran crested porcupine (*Hystrix brachyura*) which was 27 years and 3 months old when it died in 1965, at the National Zoological Park, Washington, DC. Although wild porcupines would probably not survive as long, there is little doubt that they are one of the longest-lived groups.

Rodents occupy a relatively low place in the food pyramid, with a huge number of other animals, ranging from frogs and snakes to birds of prey and cats, preying upon them. Their high reproductive rate is necessary to ensure their survival in a hostile world, as is their early age of maturity in most cases.

Chapter 4
Evolution and Distribution

Soon after the era of the dinosaurs had drawn to a close, the ancestors of today's rodents evolved, about 60 million years ago during the Late Palaeocene epoch. Their earliest fossilized remains have been unearthed in North America and indicate that the first rodents were probably similar in appearance to today's squirrels.

The earliest recorded mammals developed from a particular group of reptiles, the cynodonts, during the Triassic period, approximately 160 million years earlier. Known as tricodonts, these were small creatures which probably still laid eggs but had developed the typical mammalian pattern of dentition. This was to become modified in the case of rodents.

In the subsequent Jurassic period, two new groups of mammals came into existence. Firstly, there were the multituberculates, which show remarkable similarities to rodents in the pattern of their dentition. They had characteristically sharp incisor teeth at the front of their mouth, with a gap behind. The premolar and molar teeth had broad surfaces with cusps, which were used to crush the vegetation that these animals ate. Their jaws moved up and down, rather than side to side.

The largest multituberculates were about the size of a contemporary beaver, whereas most remained no bigger than mice. It appears that this group of early mammals gradually died out, although some survived until the Eocene period, about 55 million years ago.

Their decline occurred at the same time as the rodents began to appear but there seems to have been no direct evolutionary link between the two groups. The similarity of appearance can be explained by the fact that they had the same type of life style. In common with squirrels, for example, members of the multituberculate family Ptilodontidae had exceedingly flexible ankle joints. This not only helped them to climb up branches but also allowed them to escape danger by running down a vertical tree trunk with their feet under their bodies. Further support was provided by the sharp claws which were present on their toes. A point of difference, however, was that *Ptilodus* and other multituberculates had a flexible tail, like the present-day spider monkeys (*Ateles* species), and used their tail as an additional limb, which enabled them to keep their balance in the tree tops.

The reasons underlying the extinction of the multituberculate lineage are unknown. But it does appear that these mammals died out without leaving any direct descendants. Instead, all of today's mammals, with the notable exception of the egg-laying monotremes, are believed to have evolved from the second group of mammals, known as pantotheres, which emerged during the Jurassic period. These included the

The small size of rodents meant that their corpses were unlikely to become fossilized. They were more likely to be eaten whole by scavengers.

earliest placental mammals, which ultimately gave rise to the wide range of not just rodents but all other mammals present on the planet today.

Unfortunately, significant blanks exist in the fossil history of rodents and these can be related in part to their life styles. As small creatures, they were often preyed upon by other larger mammals. Their corpses were also more likely to be scavenged and eaten whole, which destroyed their bones, so that nothing was left to develop into fossils, as was more likely to be the case with large mammals.

This, in turn, makes it difficult to form any reliable assessment of the numbers of rodents, although it appears that they soon became numerous. The frequency of finds of their teeth, as distinct from bones, shows that their numbers were growing rapidly throughout the Tertiary period.

EARLY TERTIARY RODENTS

Squirrels were the first rodents to evolve and their earliest remains were uncovered in North America. Members of the ancient family Ischyromyidae were not unlike contemporary squirrels in appearance. A typical example is *Ischyromys* itself, which grew to an overall length of about 60 cm (24 in). It had a more mouse-like head than some other members of the family but already by this stage the double incisor teeth in the upper jaw had clearly evolved.

These early squirrels probably competed with the small primates

Rodents have probably changed relatively little through their evolutionary history in some respects. In fact, the earliest rodents were probably not unlike today's squirrels. This is the grey squirrel (*Sciurus carolinensis*), a species which originated in North America – the continent where the first rodents are thought to have evolved.

which lived in the forests of North America at this time and may, ultimately, have contributed to their extinction in this part of the world. Another similar North American rodent, which bore a distinct resemblance to a modern squirrel, was *Paramys*, which probably fed in the trees, eating seeds, fruits and possibly even bark, which it stripped off with its incisor teeth.

During the succeeding Oligocene epoch, there were significant changes in the mammalian fauna. In the northern hemisphere, approximately one third of the 95 families which then existed died out during this period. The development of the rodent families began to gain momentum, however, with pocket gophers (*Geomyidae*) starting to appear in North America.

The origins of beavers can also be traced back to the Oligocene, while the first modern genus of squirrels appeared as this epoch drew to a close. *Sciurus* is a tree-dwelling genus but the diversity in the appearance of squirrels was greatly enhanced at this stage by the emergence of both flying squirrels and species which lived entirely on the ground. This increasing diversity may have been triggered by climatic changes which led to the development of grassland areas.

Rodents also moved further south, pushing into South America during the early Oligocene. These were members of the caviomorph group, such as *Platypittamys*, which had a particularly small infraorbital foramen, with no evidence that the anterior part of the masseter muscle passed through it, as it does in later caviomorph rodents.

It is impossible to know whether the early rodents had plain coats or were marked like this African striped grass mouse (*Rhabdomys pumilio*), but it is unlikely, in view of their life styles, that they differed significantly from contemporary species in terms of external appearance.

Dormice are an ancient group of rodents, already apparently well established during the Oligocene epoch, between 36 and 24 million years ago. This is the garden dormouse (*Eliomys quercinus*), one of the most widely distributed species today, found in Europe, North Africa and western Asia.

The very number and diversity of caviomorph families in South America, however, suggest that they may have arrived even earlier, possibly during the preceding Eocene epoch. Representatives of the American porcupines, as well as pacas, spiny rats and chinchillas, and the extinct Eocardiidae family, among others, have been discovered in this continent.

It seems most likely that the ancestral caviomorphs were descended from the Holarctic family Paramyidae found in North America, Eurasia and Africa north of the Sahara, although it has also been suggested that they were carried across the South Atlantic from western Africa at a time when the ocean was much narrower. In this case they would be related to the African phiomyids, although most palaeontologists tend to dismiss this as unlikely.

During the Oligocene, diversification was also taking place in the rodent population of Eurasia. Dormice were well represented in Europe and primitive hamsters have also been discovered from this epoch. Some mice and rats were also in existence by this time and the remains of both birch mice and bamboo rats have been unearthed in this part of the world. Rodents were one of the few mammalian orders to emerge at the end of the Oligocene epoch, having radiated further afield and increased in diversity during this period of time.

MIOCENE RODENTS

Significant geological changes took place during the Miocene, with Africa moving into contact with Eurasia and the Mediterranean Sea coming into existence. The temperature across the globe continued to fall, although it was still significantly warmer than it is today. Nevertheless, this had effects on the movements of mammals. Although the Bering Land Bridge, which connected present-day Alaska with Asia, was still intact, the number of animals crossing by this route probably became less as the climate became colder.

While *Paramys* finally died out during this epoch, other established groups of rodents flourished. More than ten different genera of dormice could be found in Europe while beavers, which had first appeared during the preceding Oligocene epoch, also became more numerous. As today, they occurred in both North America and Europe.

It appears that beavers originally arose as terrestrial rodents and some may even have lived in underground burrows. In what is now the state of Nebraska, the primitive beaver *Palaeocastor* inhabited vertical tunnels, which were 2.5 m (8¼ ft) deep and have been christened 'Devil's Corkscrews' because of their distinctive shape. Another Miocene member of the beaver clan was *Steneofiber*, which lived in Europe. It was smaller than today's beavers, with less powerful teeth, and this would probably have precluded it from damming water courses.

One of the strangest groups of rodents, the horned gophers (Mylagaulidae), lived in North America for a period during the Miocene epoch. *Epigaulus*, a member of this family, in some respects resembled a contemporary marmot in terms of appearance, with claws on its front feet which indicate that it burrowed. On its nose, however, it had a very stout pair of horns, making it the only double-horned rodent ever known to have existed. The precise purpose of these horns is unclear. They may have been used for fighting or may have assisted with tunnelling in some way. Growing to about 30 cm (12 in) long, *Epigaulus* died out before the end of the Miocene, as its forest home gave way to grassland. Other, similar mylagaulids were equipped with a single horn. None of these horned gophers survived into the Pliocene epoch.

Nevertheless there were some rodents from the Miocene era whose descendants are still present in North America today. During the Miocene, it appears that members of the family Aplodontidae were numerous. Today, this group is represented by a single species – the mountain beaver, or sewellel (*Aplodontia rufa*), which, in spite of its name, is not a beaver in the true sense, although it can swim well if necessary.

Further south, there was an increase in the size of some rodents, a trend which culminated in the emergence of the Dinomyidae, which gave rise to the present-day pacaranas. This family included the largest rodents that ever existed and, like the guinea pig family (Caviidae), it evolved and rapidly diversified during the Miocene.

Climatic changes probably triggered the rapid evolution of a number of these forms. *Eocardia* grew to about 30 cm (12 in) in length, with longish legs. *Telicomys* was the giant of this family, reaching 2 m (6½ ft) in length, equivalent in size to a hippopotamus. These dinomyids, whose name means 'terrible mice', had stout tails and a long, broad-snouted head. They survived until the early Pliocene but unfortunately little is known about them.

Another large rodent from South America, which lived during the Miocene, was *Protohydrochoerus*. This grew to about the size of a tapir but, again, was almost certainly primarily vegetarian in its feeding habits.

PLIOCENE RODENTS

The Pliocene epoch was of relatively short duration, lasting for about 7 million years. In geological terms, it was a time of relative stability. The shape of the North American continent was similar to that of today but, in Europe, Britain became connected to the rest of the continent, as sea levels receded. The relative shortage of rain encouraged the spread of grassland areas, particularly in the Americas. Asia also suffered from the effects of drought, and the world was generally warmer than today, with Siberia, for example, having a temperate climate.

Fossil evidence reveals that many rodents fell victim to hunting birds, notably owls, just as they do today. Dormice (Gliridae, Seleviniidae) were in decline, although the origins of contemporary genera can be traced back to this period of time. It was also in Europe that the bizarre naked mole rat (*Heterocephalys glaber*) first arose while, further east, the steppelands of Asia provided an ideal habitat for jerboas (Dipodidae).

Modern dormice genera, such as *Eliomys*, had also developed in the Pliocene epoch, although the number of species was apparently starting to decline at that time.

Bamboo rats (*Rhyzomys* species) also established themselves in Asia, before invading Africa during the subsequent Pleistocene epoch.

Porcupines were already represented throughout the Old World by this stage and were probably quite distinct from their New World counterparts, which originated in South America.

Significant changes started in the Early Pliocene with the emergence of the first rats (*Rattus* species) and mice (*Mus* species). Voles (Microtinae) also became more numerous. The development of both these groups continued apace throughout the Pleistocene. By this time, squirrels (Sciuromorpha) were already widely distributed and had evolved into a number of forms. Some ground-dwelling groups, notably marmots (*Marmota* species) and prairie dogs (*Cynomys* species), remained confined to North America.

Beavers, including the *Castor* genus represented in both North America and Eurasia today, had become well established, although *Castor* itself did not emerge until the Early Pliocene. These particular beavers fed on bark, whereas other genera at that time were browsers on water plants.

During the Pliocene, South America again became linked to the northern land mass. This enabled mice to spread south but very little is known about them from the fossil record. The significance of possible movement between Africa and South America via the sea, with mice travelling on wood or other debris, is unclear. Bearing in mind the high metabolic rate of these creatures, it seems unlikely that they could have survived a sea crossing in any numbers without ready availability of fresh

One of many mysteries concerning the evolution of rodents is whether there was any direct link between species occurring in Africa, such as spiny mice (*Acomys* species), and those in South America, at times when Pacific Ocean was narrower than it is today.

food and water. The journey would have required them to travel several thousand kilometres at the mercy of the ocean currents rather than in a straight line.

Rafting may well have been more significant, however, in explaining the early presence of mice in the West Indies, as well as on the volcanic Galapagos Islands. This would have involved a much shorter crossing, although it is unlikely that small rodents would be well adapted to this means of spread, compared, for example, with tortoises, which also invaded the Galapagos Islands from the Americas.

It seems likely that mice of the subfamily Sigmodontinae evolved in South America rather than invading from the north. Support for this view comes from a study of their ectoparasites, notably fleas and mites, which show a much closer affinity to those found in South America or Australia than those which occur further north. Chromosomal research indicated that these mice probably originated in northwest South America and then spread south across the continent. In typical rodent fashion, these mice subsequently adapted to virtually every niche available to them, developing into 40 genera comprising approximately 200 species.

A number of North American families of rodents, such as the pocket gophers (Geomyidae) and spiny pocket mice (Heteromyidae) clearly

Old World porcupines, forming part of the caviomorph grouping, had evolved by the Pliocene epoch, about 5 million years ago. This is the African porcupine (*Hystrix cristata*).

had difficulty in penetrating successfully southwards into the southern continent. This was probably because of differences in the vegetation, notably the presence of rainforest rather than grassland.

PLEISTOCENE RODENTS

The cooling climate which resulted from the increasing development of mountains around the world gave way to an ice age during the Pleistocene. The effects of this were most severe in the northern hemisphere, where much of North America and northern Europe were buried under sheets of ice. To the south of this line there was an area of tundra, not unlike that seen in Arctic regions today. The Mediterranean region was decidedly cooler than it is at present and supported forests of pine trees.

During the warmer interglacial periods, the ice sheets started to retreat, enabling hardy trees, such as birch, to recolonize the tundra region. At the southern end of the world, Antarctica was covered in ice throughout the Pleistocene, as it is today. (This area began to have a permanent ice covering in the Miocene epoch, over 20 million years earlier.)

The level of the oceans fell, re-creating the Bering land bridge between Asia and northwestern North America. This allowed hardy species again to cross from one continent to another. The English Channel was also dry, enabling animals to move back and forth from the Continent.

The Gulf of Sunda shelf, which supported various islands, e.g. Java, Borneo and Sumatra, off the southeast coast of Asia, also dried up. These islands then became part of the continuous land mass of Asia, although the route to New Guinea, and Australia beyond, still remained blocked by deep sea.

As cold periods and warmer intervals occurred throughout the Pleistocene, so opportunistic groups were able to exploit these climatic shifts. Rats (*Rattus* species) developed rapidly, particularly during the latter part of this epoch, while voles also thrived. Increasing signs of specialism can be seen, with members of the Microtinae, such as lemmings, becoming firmly established in tundra regions. They entered North America, while marmots (*Marmota* species) spread to Eurasia.

The beginnings of the typical present-day circumpolar fauna started as the continents became linked again and the habitat around the northern part of the world became similar. In voles (Microtinae), the development of rootless cheek teeth, growing throughout life, also occurred, marking a significant advance in the feeding potential of this group.

An unusual occurrence took place on some of the islands in the Mediterranean Sea. At this time, they supported a much greater range of mammals than today. Elephants, in the company of hippopotamuses, roamed there but, in order to survive in a limited area of land, they became smaller in size.

Strangely, the reverse happened in the rodents which shared these islands with them. Among these large rodents was a giant dormouse (*Leithia*), which lived on a number of Mediterranean islands. In terms of appearance, it was basically indistinguishable from the *Muscardinus* dormice which now inhabit Europe but was significantly larger, measuring 25 cm (10 in) long, excluding its tail.

The increase in size of these dormice may have arisen simply because they faced no predators or it may have been a means of population control. Larger animals require more food and so are likely to be less numerous in a given area than their smaller counterparts. In the case of rodents, with their prolific breeding habits, the risk of the population rapidly outstripping the available food supply would be increased, but larger animals tend to mature more slowly and have fewer offspring than their smaller counterparts.

Unfortunately, nothing is known about the biology of these giant dormice but they were not unique. Although there is a possibility that the giant dormice on adjoining islands, such as Malta and Sicily, were closely related, it is harder to explain their presence on more widely separated Mediterranean islands. They probably evolved in isolation from populations which originally shared a common ancestry.

Across the Atlantic Ocean in North America, another giant form of rodent arose during the Pleistocene epoch. The giant beaver (*Castoroides chicensis*) ranged south from Alaska to Florida and east from Nebraska. Its major stronghold appears to have been south of the Great Lakes. Growing to 90 cm (36 in) at the shoulder and weighing perhaps 90 kg (198 lb) – which is about the size of an American black bear (*Ursus americanus*) – the giant beaver lived in swampy areas, where it fed on coarse vegetation. Unlike contemporary beavers, it did not appear to build dams, nor did it lead a primarily aquatic existence. Its front legs and feet were small in contrast to its large hind limbs, which probably had webbed feet. Changes in habitat may have led to its ultimate extinction. There is no evidence that it was hunted by early people in North America. Another large form of beaver, *Trogontherium*, arose in Eurasia at this stage, although it was nowhere near as large as *Castoroides*. These animals appear to have been more aquatic.

A number of rodents from South and Central America pushed northwards during the Pleistocene, including two forms of the capybara, which extended as far north as South Carolina. One species, *Hydrochaeris holmesi*, lived in Florida and was similar in size to those seen today, although its cheek teeth were bigger, while the other, the giant capybara (*Neochoerus pinckneyi*), was nearly twice as large. Its remains have been found in Texas and South Carolina, as well as South America, suggesting that it may have had quite a wide distribution.

Another invader from further south was the short-tailed porcupine (*Erethizon dorsatum*), which fared rather better than the capybaras. It became widespread throughout North America during the Pleistocene epoch and still retains this position today, being found from Mexico north to Alaska.

The Pleistocene marked the beginning of a new trend which has continued to exert an effect on the distribution and populations of rodents around the world. The emergence of modern man (*Homo sapiens*) has had far-reaching consequences for rodents in both positive and negative senses; perhaps more so than for any other group of mammals. Although very rarely welcomed, rodents have lived in human company for millennia. A number of species have benefited from this association but others have been exterminated, particularly those which had a limited distribution, typically the island forms.

DIVERSIFICATION OF RODENTS

The Caribbean

The Caribbean region is the site of the greatest number of rodent extinctions since the Pleistocene and this is due in part to human pressures.

The heptaxodont rodents formed a family which was once widely distributed throughout the West Indies and some certainly survived until the sixteenth century. Their remains have been found on a number of islands, including Jamaica, Puerto Rico, Haiti and the Dominican Republic, although the precise number of species involved is unclear.

The name 'heptaxodont' refers to their cheek teeth, which could have up to seven laminae and had parallel crests running across the teeth. They had a characteristically stocky appearance and a huge skull.

The attractive Cuban hutias (*Capromys* species) are some of the survivors of the rodent fauna of the Caribbean region, where more extinctions have occurred over the past 400 years than anywhere else in the world.

Of these heptaxodonts, the Hispaniolan quemi (*Quemisia gravis*) was reportedly about the size of a paca (*Agouti paca*) and brown in colour. Like the Puerto Rican species (*Elasmodontomys obliquus*), which died out perhaps 200 years earlier, in about 1500, the quemi was heavily hunted by the native people for food. This, and the introduction of ship's rats, which may have brought unfamiliar diseases, contributed to the demise of these rodents.

The third heptaxodont, *Heptaxodon bidens*, which survived until modern times, lived on Puerto Rico until about 1600. It was about the same size as a small hog, growing to an overall length of about 30 cm (12 in). Findings of its skeletal remains are very limited but it was clearly widely distributed across the island until the time of European settlement. Changes in habitat and hunting pressures probably led to its extinction.

Other rodents on Puerto Rico were also valued as food. It appears that the Puerto Rican hutia (*Isolobodon portoricensis*) may even have been kept and bred by the islanders for a period. A related species, *Isolobodon levir*, occurred on Haiti where its remains have been found in middens, as well as in caves which were home to the giant barn owl (*Tyto ostalaga*). This Haitian (Hispaniolan) hutia died out at a relatively early stage in history, just after European settlement had begun. The natives then began to rely more heavily on the Puerto Rican species, as well as domestic guinea pigs, as sources of food.

Other rodents also occurred on Puerto Rico at this time but even less is known about them. It is believed that there were two different types of agouti, which may have been similar to those still living on Trinidad

Hutias (*Capromys* species) are a group of rodents confined to the Caribbean region, where they were apparently far more diverse in the past. How they reached these islands is unclear.

today, although smaller in size. They had died out by the middle of the eighteenth century.

The Caribbean islands were also the home of a number of different species of spiny rats (*Brotomys* species). Two species that formerly occurred on Hispaniola, and a further couple from Cuba, have disappeared in recent times. Smaller than the native hutia, but a similar shade of grey, the Hispaniolan spiny rats had a coarse, erect coat, which was probably not unlike that of the spiny mice (*Acomys* species) of today. The Hispaniolan species, *Brotomys contractus* and *B. voratus*, could apparently be distinguished by their particularly narrow palate. Both were considered delicacies among the island chiefs and this may have led to their extinction in about 1600. The Cuban spiny rats survived longer, until the end of the 1860s.

The introduction of mongooses through much of the Caribbean proved fatal for a number of the indigenous rodents. This early example of biological control was carried out primarily to curb the numbers of black rats (*Rattus rattus*) and brown rats (*R. norvegicus*), brought from Europe, which had multiplied freely and were proving to be a major pest in the sugar-cane fields.

By 1789 it was estimated that these rats were destroying about a quarter of the entire crop on Jamaica alone. Unfortunately, the arrival of the mongoose soon sealed the fate of the Jamaican rice rat (*Oryzomys antillarum*), which was unusual in this part of the world, in being a cricetine rather than a caviomorph rodent. It had become extinct by the 1880s. Described as having a reddish coat, with a milk-white belly, the Jamaican

Some rodents are predatory, while the adaptability of others means that they can pose a threat to indigenous species. The forest dormouse (*Dryomys nitedula*) is known to eat other smaller rodents on occasions, as well as birds.

rice rat lived in burrows along river banks and in hollows under large trees. A similar but slightly smaller species occurred on the island of St Vincent, measuring about 20 cm (8 in) overall, with its tail accounting for approximately half of this total. It, too, vanished, thanks to the introduction of mongooses in about 1900; only one specimen is known to science.

Huge muskrats (*Megalomys* species), which could grow almost as large as cats, were a unique group of rodents endemic to the Lesser Antilles in the Caribbean. There were probably three species in total, with the Barbuda muskrat (*M. audreyae*) being the first to become extinct in about 1600. These distinctive black-and-white rats had a characteristic musky odour. In order to remove this, the natives singed their fur and boiled them before eating them. The St Lucian muskrat (*M. luciae*) finally died out in about 1880, with the Martinique species (*M. desmarestii*) vanishing after a severe volcanic eruption in 1902 unleashed clouds of poisonous gas into the atmosphere.

Australasia

The original mammalian fauna of Australia consisted of the monotremes, or egg-laying mammals, which probably evolved in this part of the world and are still represented today by the duck-billed platypus (*Ornithorhynchus anatinus*) and the two species of echidna comprising the family Tachyglossidae. The arrival of the placental mammals in Australia occurred relatively late in the geological record. Bats were probably the first group to arrive, perhaps as early as the Eocene, followed by rodents during the Pliocene.

It is likely that human settlement in South-East Asia and Australia has assisted the spread of rodents, as it has in other parts of the world. They can be inadvertently transported in ships' cargoes and thus distributed among the many thousands of intervening islands.

Nevertheless there are peculiar difficulties facing the spread of mammals in this part of the world. Zoologists have identified a barrier that appears to separate Oriental and Australian faunas and which is known as Wallace's Line, after the Victorian explorer and naturalist, Alfred Wallace. This barrier cuts between Borneo and Sulawesi, where there is a gap of only 24 km (15 miles), and continues between Java and Bali. However, this barrier has not been consistent in the geological past. New Guinea, for example, was closer to Borneo in the Pliocene epoch, which would have made it much easier for rodents to move between these islands.

It has been proposed that there were seven separate periods in the geological past during which rodents entered Australia and these probably began in the Pliocene. Evidence for the route which they took can be gathered from a comparison of present-day distributions of genera. Water rats (Hydromyinae) lacking posterior molar teeth and with distinctive cheek teeth, are confined mainly to New Guinea, where nine genera are to be found, while one has established itself successfully in Australia.

The Hastings River mouse (*Pseudomys australis*) is a small rodent, found in parts of Queensland, which was thought to have become extinct, but has recently been rediscovered. *Pseudomys* mice are indigenous to Australia, with about 17 species being recognized, some of which have a very restricted area of distribution.

Other species, notably murids, clearly spread and adapted to a variety of evolutionary niches after reaching Australia. Many have since developed to survive in the inhospitable conditions at the centre of the continent, where water is often in short supply. While both rats and mice are well established there, the absence of squirrels is perhaps surprising. This is perhaps a reflection of the effectiveness of Wallace's Line as a barrier. The distribution of squirrels was curtailed on Sulawesi, where they are joined by a single species of porcupine.

The diversity of rodents in Australia is therefore less than that in other continents. Those like water rats are sometimes described as 'Old Papuan forms' because their distribution is centred in that part of the world. Relatively few have crossed into Australia.

In contrast, the Pseudomyine, the 'generalist' rats and mice, have radiated considerably since their arrival in Australia. They have probably benefited most of all from human settlement of this continent.

Another group has evolved into relatively large, tree-dwelling rats (*Mesembriomys* species), which fill the same evolutionary niche as squirrels elsewhere. The desert has encouraged the development of rodents which progress by jumping, rather like jerboas, with correspondingly large ears to pick up and isolate sounds in open terrain. These are the Australian hopping mice (*Notomys* species). A group of Muridae has evolved a vole-like appearance, featuring a relatively short tail and slender feet.

Finally, there are two species of rats (*Leporillus* species) which build stick nests. Unlike the other groups, which have clear parallels in Eurasia and elsewhere, the stick-nest rats have no obvious counterparts, especially in view of their highly social natures. Their localized area of distribution indicates that they may represent a rather aberrant and vulnerable group of rodents.

Although rodents managed to cross successfully to Australia, none made the onward journey to New Zealand. In fact, in the geological past, there were no mammals there at all. Instead, bizarre and sometimes huge birds evolved to dominate the landscape.

The first definite report of rats being brought to New Zealand dates back to 1350. The Polynesian settlers who travelled there by canoe brought various native foods with them, as well as the Pacific rat (*Rattus exulans*), which became known as the *kiore*. Legend tells how the first rats may have been carried in the war canoe of Te Kupe, a famous chief. It has been suggested, however, that these rats were introduced to Chatham Island by the Moriori people, who preceded the Maori settlers.

These rats multiplied readily in New Zealand, although whether they caused any damage to the native fauna is unclear. Certainly they were heavily trapped for food by the Maoris on both North and South Island. Pacific rats eat more fruit than European rats, so that the environmental impact of their introduction was probably less damaging.

Certainly, once the first vessels started to arrive from Europe, bringing brown rats (*Rattus norvegicus*) with them, the escape of these rodents apparently spelt disaster for the kiore. Charles Darwin, during his

While the house mouse (*Mus musculus*) has spread far beyond its natural range, thanks in part to inadvertent human assistance, other species, such as the African striped grass mouse (*Rhabdomys pumilio*), remain localized in particular habitats.

voyage on HMS *Beagle*, visited New Zealand in 1835 and recorded how the new invaders were said to have eliminated the more docile kiore from the northern part of the island in barely 2 years.

Kiores managed to survive in central areas, particularly around Rotorua, and reached plague proportions between 1841 and 1842 in the vicinity of New Plymouth, surprising European settlers with their climbing skills.

A further challenge to the survival of the kiore in New Zealand occurred with the introduction of a second species of rat – the black rat (*Rattus rattus*) – from Europe. This took place during the 1850s, since when kiores have become progressively scarcer, almost dying out by the turn of the century.

The population on South Island fared somewhat better, building up a probable peak in 1884. The kiore, although now extremely scarce, is still found in the Fiordland area and on offshore islands, where neither of the two introduced species of rat is present.

The part played by the kiore in shaping the avifauna of New Zealand in the centuries since its introduction is unclear. It may have hastened the extinction of the moas, which include the tallest known birds ever to have existed, at a time when they were being heavily hunted. These rats would certainly have preyed upon the eggs and fledglings of these ground-dwelling birds. In more recent times, they have been blamed for the extinction of the pied tit (*Petroica macrocephala toitoi*) on Cuvier Island and the yellow-breasted tit (*P. m. macrocephala*), which used to inhabit Inner Chetwode.

Pacific rats are less adaptable than their European relatives to the habitat changes caused by agriculture. This probably contributed to their demise, as did the introduction of mustelid predators, notably ferrets, stoats and weasels, which were originally brought to New Zealand to control the burgeoning rabbit population. Rabbits were introduced from England in the early 1860s and multiplied rapidly.

The third species of rodent introduced to New Zealand as the result of European settlement was the house mouse (*Mus musculus*), although the date of its arrival is unclear. These mice probably travelled undetected in cargoes of grain and escaped ashore at different localities from the early 1820s onwards. Some were also shipwrecked and managed to swim ashore. House mice reached the island of Ruapuke, in the Foveaux Straits, after the *Elizabeth Henrietta* foundered off the shore in 1824. The local people called them 'henriettas' – a term which persisted for more than half a century after their first appearance.

A similar story applies to the introduction of the house mouse to Australia. Brought in cargoes on the ships which carried the early settlers, these mice soon built up to huge numbers as wheat started to be grown in the fertile plains of the continent. House mice were soon causing massive damage, in spite of the fact that many millions were being killed regularly with poison.

Chapter 5
The Order Rodentia

The fact that nearly 40 per cent of the world's mammals are rodents has posed particular problems for taxonomists, whose task is to classify both living and extinct organisms. In the case of rodents which range over a wide area, it can be especially difficult to know when one form diverges sufficiently from another to be recognized as a distinct species.

There was once a tendency to divide populations at every opportunity but, today, a more pragmatic approach is used. New means of determining relationships between groups of rodents, notably the introduction of DNA (deoxyribonucleic acid) testing, have facilitated taxonomic studies and the impact is likely to revolutionize our future understanding of rodent relationships.

DNA is arranged in strands which take the form of a double helix. There are present in every living cell in the body. Within these strands, sequences of pairs of acids and bases are built up. Strands of the DNA from a number of animals will reveal how many of these sequences they have in common and this enables an assessment to be made of how far they have diverged from each other.

As the genetic material of the organism, DNA can also be used in an historical sense, since it is passed on from one generation to the next. Scientists can therefore use DNA analysis to estimate at what point in the past related groups split from each other. This is likely to be particularly significant in the case of rodents, where the fossil evidence of their evolution and development is very limited.

The classificatory system is tree-like, consisting of a series of progressively less generalized ranks, which culminate in the subspecies. Taxonomic disagreements tend to be at this level, where there is frequently dispute about the validity of particular subspecies.

Within the rodent order, there are three major groupings: the sciuromorphs, myomorphs and caviomorphs. At present, below this level, taxonomists generally recognize 30 families, 389 genera and 1,702 species. The following example show the classification of the mountain thicket rat, an African rodent:

Order:	*Rodentia*
Suborder:	*Myomorpha*
Family:	*Muridae*
Subfamily:	*Murinae*
Genus:	*Thamnomys*
Species:	*Thamnomys venustus*
Subspecies:	*Thamnomys venustus venustus*
	Thamnomys venustus kempi

The first subspecies to be described, known as the nominate subspecies, can be identified by the repetition of the specific epithet, which is *venustus* in the above example. *Thamnomys venustus* was originally described in the scientific literature in 1907 and *Thamnomys venustus kempi* was identified later in the same year. It was at first thought to be a separate species.

New rodents are still being discovered. Their generally small size and often limited area of distribution, particularly in the case of rainforest rodents, mean that they can escape detection.

One of the most amazing discoveries of recent years has been that of the pacarana or Branicki's terrible mouse (*Dinomys branickii*). It was originally described by Professor Constantin Jelski in 1873, when he was in South America on an expedition financed by Count Branicki. Early one morning Professor Jelski spotted one of these huge rodents, whose body can be nearly 90 cm (36 in) long overall, in an orchard. He was able to kill it and sent its remains back to Poland.

Studies revealed that the pacarana shared some features with the pacas (*Agouti* species), notably in terms of size and coat. In fact, *pacarana* is a native Tupi Indian word meaning 'false paca'. However, there were also significant differences between these rodents, not the least of which was the fact that the paca had five toes on each foot, whereas the pacarana had just four.

Its tail, which accounted for approximately one quarter of its total body length, and body profile were further points of distinction, while the shape of its molar teeth suggested a closer affinity with the chinchillas (*Chinchilla* species). In fact, the pacarana displayed features drawn from a number of other rodent groups. It was so different that it was finally grouped in a separate family, Dinomyidae, of its own.

Subsequently, zoologists feared that this unique gigantic rodent might have died out. For over 30 years nothing further about it was recorded. Then, in 1904, two living pacaranas were sent to the Para Museum in Brazil, having been obtained from the upper reaches of the Rio Purus, close to the Peruvian border.

Although Jelski had christened this species 'Branicki's terrible mouse', presumably on the basis of its size, Dr Goeldi, who received the living pacaranas, found that they were in no way terrifying. In fact they proved to be friendly and peaceful, as well as quiet, although they did growl and hiss on occasions. These pacaranas were kept in the Zoological Gardens in Para, Brazil, where the female gave birth, although complications developed and she died soon afterwards.

The taxonomic problems which can arise were then highlighted by the discovery of another pacarana in Colombia, which, for a time, was recognized as a separate species, *Dinomys gigas*. A greyish black rather than brown pacarana was also accorded specific status, as *Dinomys pacarana*, in 1919. Further pacaranas were then obtained from parts of Brazil and Peru.

The confusion reigned until 1931, when Dr Colin Sanborn published a review of the genus, based upon the first detailed study of the various specimens. This study confirmed that there was in fact only one species, *Dinomys branickii*, and that the other two supposed species of *Dinomys* were simply variants which did not merit separate recognition.

Another taxonomic oddity from South America whose history is similar to that of the pacarana is the thin-spined porcupine (*Chaetomys subspinosus*). This appears to be a link between the American porcupines and the spiny rats (*Proechimys* species), possessing a tail which is just partially prehensile, with a covering of stiff spines.

Thin-spined porcupines are confined to southeast Brazil. First recorded in 1818, they appeared to be very rare and it was thought possible that they might have become extinct. In fact, from 1952 until December 1986, there was no confirmed sighting of them.

Even in restricted areas it can be hard to locate rodents on occasions. Rice rats (*Oryzomys* species) are widely distributed in the New World and some have managed to establish themselves on the Galapagos Islands, off the western coast of South America. Populations there have suffered following the introduction of rats (*Rattus* species). It was feared that the San Salvador Island rice rat (*O. swarthi*) had become extinct, after being last seen in 1906, but the find of a fresh skull in 1966 confirmed the continuing existence of this species.

The status of a surprisingly high number of the world's rodents is unknown at the present time. Almost certainly a number of them are likely to be in danger of extinction. Some are already recognized as being endangered, like Cabrera's hutia (*Capromys angelcabrerai*), which was only discovered in 1974. This hutia lives in the mangrove swamps of Cayos de Ana Maria, in south-central Cuba. As a group, the Cuban hutias are probably the most endangered rodents in the world. One of their number, the dwarf hutia (*C. nana*) has not been officially recorded since 1937 but is believed still to survive in the Zapata Swamp, which is now a protected area.

Rodents which appear to have very specific environmental requirements, such as the San Quentin kangaroo rat (*Dipodomys gravipes*), are especially vulnerable. This North American species inhabits a flat plain, with low vegetation, which extends over a distance of only 100 km (63 miles) along the coast of northern Baja California, Mexico. A change in habitat occurred during the 1970s when most of the area was ploughed up. From being very numerous throughout the region in 1972, the San Quentin kangaroo rat had apparently largely disappeared by 1980, as the direct result of this habitat modification.

Work to conserve the world's endangered rodents is coordinated mainly by the International Union for the Conservation of Nature and Natural Resources Species Survival Commission (IUCN–SSC) Rodent Specialist Group, which was set up in March 1980.

Unfortunately, the subject of rodent conservation is not one which proves to be generally popular with the media, which are far more disposed to show sympathy for endangered megafauna, such as the tiger (*Panthera tigris*). However, recent campaigning in the UK to save the native dormouse population indicates that people are not ill-disposed towards rodents in general. This does give some cause for optimism for the future of those species whose existence is currently threatened.

Chapter 6
Sciuromorpha: Squirrel-like Rodents

This group of rodents consists of seven different families, with an almost worldwide distribution. Sciuromorphs are present on all continents apart from Australia and Antarctica. Many squirrels are arboreal and diurnal by nature. The sciuromorphs also include the two families of beaver (Aplodontidae and Castoridae), the pocket gophers (Geomyidae), pocket mice (Heteromyidae) and the springhare, which is the sole member of the family Pedetidae.

MOUNTAIN BEAVER • FAMILY APLODONTIDAE
The sole member of this family is found in areas of coniferous forest on the western seaboard of North America. The mountain beaver, or sewellel (*Aplodontia rufa*), is mainly nocturnal in habit and spends much of its life underground, where it excavates a series of tunnels and interconnecting chambers. It generally lives alone, moving below the ground in tunnels which may be just 15 cm (6 in) below the surface. Its food stores and nesting chambers are at a much greater depth, perhaps 2 m (6½ ft) underground. Other animals, including salamanders and even other rodents, notably deer mice (*Peromyscus* species), may be found in these burrows.

Food stores in the mountain beaver's nest are especially important during the winter, since it does not hibernate, partly, it seems, because its metabolism does not allow it to build up fat stores. It also has a high water requirement.

The mountain beaver is exclusively vegetarian in its feeding habits, eating a wide range of greenstuff and twigs. Plants with a high moisture content are likely to be stored underground. Sword fern fronds are particular favourites but, when food supplies at ground level are short, the mountain beaver is able to climb trees, to a height of about 7 m (23 ft). It nips off branches with its sharp incisors as it climbs up the tree and, although it may drop some to the ground, it more usually climbs down headfirst with the branches in its mouth. These rodents can cause severe damage in forestry plantations and the cost of their attacks may amount to millions of US dollars annually.

In its burrows the mountain beaver has special faecal chambers where it secretes its droppings, some of which it ingests to obtain the maximum nutritional benefit. These primitive rodents are slow to mature and slow to breed. Mating takes place early in the year and the breeding season is quite short. Females produce a single litter between February and April.

Distribution of squirrel-like rodents (Sciuromorpha).

Pregnancy last approximately 4 weeks and most litters consist of just two or three offspring, although four have been recorded occasionally. Helpless at birth, young mountain beavers develop slowly and their eyes are not fully opened until they are about 7 weeks old. They are suckled for the first 2 months and remain in their mother's burrow until the end of the summer.

Subsequently they move on to find their own burrows. At this stage, when they are above ground, they are especially at risk from predators such as birds of prey and coyotes (*Canis latrans*).

Their requirements for damp surroundings and forested land mean that the young tend not to disperse very far afield, moving a maximum of 2 km (1¼ miles). In some cases, they may take over an abandoned network of tunnels near their mother's burrow. As long as the mountain beaver stays above ground, it is vulnerable.

Assuming that they survive this critical early stage in their lives, mountain beavers may live for 6 years. They are unlikely to mature until their second year. In spite of their name, these rodents are not found at high altitudes. They range up to about 2,200 m (7,300 ft) above sea level but are frequently found at much lower altitudes. Nor are they highly aquatic like true beavers (*Castor* species), although they can swim if necessary.

The range of the mountain beaver does not appear to have altered significantly over recent centuries. In fact, forestry management practices, notably the clearing and replanting of stretches of coniferous woodland, may have enabled them to live at greater densities than in the past. In favourable habitat, particularly where forests are regenerating, the density of mountain beavers can exceed 20 animals per hectare (2½ acres).

SQUIRRELS • FAMILY SCIURIDAE

This family, comprising 49 genera and approximately 267 species, is the largest in the suborder Sciuromorpha. It embraces not only the true squirrels, but also chipmunks, marmots and prairie dogs. Squirrels are among the most conspicuous of rodents, often having bold and active natures.

There is considerable diversity in size within this family, which ranges from the tiny African pygmy squirrel (*Myosciurus pumilio*), weighing under 10 g (⅜ oz), to the terrestrial marmots (*Marmota* species), which can be as heavy as 7.5 kg (16½ lb). The giant squirrels (*Ratufa* species) originating from southern Asia are the largest arboreal group in terms of size and can weigh up to 3 kg (6½ lb). They spend almost their entire lives in the tree tops, building large nests of twigs for breeding purposes.

Giant squirrels support themselves by using their tail as a counterbalance when feeding. They do not curve their tail over their back but drape it over one side of the branch, with the head and front part of their body leaning on the opposite side. These squirrels use their hind limbs to grip the branch, holding food in their front paws, with the aid of their broad, short thumbs.

Female giant squirrels produce several litters each year, consisting of between one and three youngsters. Pregnancy lasts for about 4 weeks. Their life span, at least in captivity, may be over 20 years. They can jump and move quickly through the forests where they live, leaping up to 6 m (20 ft) in a single bound. Like other arboreal species, giant squirrels can judge distances very accurately.

When it comes to gathering food being small can have advantages in the tree tops. The African pygmy squirrel is able to run over broad

Distribution of squirrels (Sciuridae).

Flying squirrels are found in many parts of the world. In spite of their name, they glide, thanks to membranes running along the sides of their bodies, rather than fly.

branches, where it digs out pieces of bark with its sharp incisors. The coloration of this species is unusual: much of the body is a greenish buff and the underparts are olive-white.

Similar in size is the tiny Asiatic pygmy squirrel (*Exilisciurus* species), found on the island of Borneo and in the Philippines. These squirrels may be no longer than 11 cm (4¼ in), including the tail, and weigh just 15 g (½ oz). Apart from feeding on vegetation, they also seem to consume ants, which they find either on the ground or in the trees.

Movement from tree to tree can be important, not only to escape predators but also to obtain food more easily. Although squirrels are able to jump long distances, they can travel much further by being able to glide, using folds of skin, which are normally inconspicuous, along the sides of their bodies.

A number of different genera of squirrels have developed this skill, the heaviest being the giant flying squirrels (*Petaurista* species), which can weigh up to 2.5 kg (5½ lb). Nocturnal by nature, these Asiatic squirrels will only glide if they are unable to reach a neighbouring tree by jumping. They will then scamper up to a higher vantage point and leap down, extending their front legs forwards and their hind legs backwards, stretching taut the membrane of skin at the sides of their body. They glide in a downwards direction but are able to climb slightly as they come into land on the tree. These flying squirrels are able to bank in flight if necessary and may also use air currents to increase the distance which they can travel.

Like other squirrels, they are mainly vegetarian in their feeding habits, although they may eat some insects, as well as stealing birds' eggs. Measuring up to 1.2 m (4 ft) long, including the bushy tail, which is often longer than the body, these giant gliding rodents are heavily hunted in some parts of their range, particularly in Taiwan, where they are a popular source of food. Here, in some areas, particularly in spring, giant flying squirrels may become too heavy to glide, having gorged themselves on the new growth of leaves and flowers.

In total, there are approximately 35 different species of flying squirrel, with representatives in North and Central America as well as the Old World, although none has reached South America. The most southerly of the two New World species is *Glaucomys volans*, whose range extends from southeastern Canada to Honduras. This flying squirrel prefers to use existing tree holes, sometimes those created by woodpeckers, for nesting sites. On occasions, they may invade the roof space of houses where they build a nest with a soft lining.

As with other species, this New World flying squirrel moves slightly upwards as it lands, by raising the tail. The front legs absorb the impact and, almost simultaneously, the squirrel will scamper around to the other side of the trunk to escape any potential aerial predator, such as an owl.

Even in the northern part of their range, these squirrels do not appear to hibernate, although they become less active when the weather is bad. As appears to be the case with other flying squirrels, litter sizes tend to average about two or three youngsters, although up to six have been recorded on occasions.

The coloration of flying squirrels may be protective, as demonstrated in the Old World species, *Pteromys volans*. Their colouring enables them

The grey squirrel (*Sciurus carolinensis*) is one of a large group of tree squirrels found in the New World. They come down to the ground quite regularly to forage for nuts and other food.

to blend in well with the bark of the coniferous trees which they inhabit throughout Eurasia, from Finland to the western seaboard of the Pacific. As a further means of escaping predators, such as owls, these squirrels rest with their bodies flattened against a tree trunk.

Undoubtedly the most bizarre of the flying squirrels is the woolly flying squirrel (*Eupetaurus cinereus*), a rare species found in the mountainous area of northern Pakistan and extending to Tibet. Its body is covered with very thick, soft woolly fur, as are its feet. This fur doubtless helps it to keep warm. Its claws are blunt rather than sharp, which suggests that this rodent inhabits rocky areas. It probably feeds on vegetation and may also hunt for insects.

While most squirrels are predominantly vegetarian, eating not just fruit and nuts but also bark, buds and leaves of trees, a few species are insectivorous. The long-nosed squirrel (*Rhinosciurus laticaudatus*), from South-East Asia, is a specialist insectivore which feeds on the ground. It hunts ants, termites, earthworms and other invertebrates, using its thin, long lower incisors and very short upper incisors to gain a secure grip on its quarry. The skull of this particular squirrel is notably elongated and the tongue is long, probably to sweep up ants. In addition to invertebrates, long-nosed squirrels also eat some fruit. Their teeth may become worn down as they become older, probably by grit ingested at the same time as their prey.

A squirrel with an even longer skull is the Celebes long-nosed squirrel (*Hyosciurus heinrichi*), which lives in the mountains on the Asiatic island of Sulawesi (formerly Celebes). Almost nothing has been recorded about the habits of this species, apart from the fact that it lives on the ground. It has a short tail and large powerful claws, which it uses for burrowing.

By way of contrast, the tree squirrels (*Sciurus* species) are some of the best-known members of the family. There is sometimes confusion between the red squirrel (*S. vulgaris*), which is distributed through Europe and northern Asia, and the red squirrels seen in North America, which belong to another genus, *Tamiasciurus*. Most species are to be found in the New World, with three being native to the Old World; none occur in Africa.

Tree squirrels are found in areas of woodland and, although they spend much of their time in the trees, they occasionally descend to the ground to gather nuts and other food. These squirrels are particularly adept at opening the hard shells of nuts with their incisors. American chestnut trees are a favourite source of food, as are oaks, whose acorns present no challenge to these tough-toothed rodents. They may bury nuts and acorns for the winter, when other foods are likely to be relatively scarce.

Tree squirrels do not hibernate but have a snug nest. They may build this themselves in the branches of a tree or they may take over a tree hollow. The young are born in the nest and are totally helpless at birth. Their fur begins to appear when they are about 14 days old and, in the early days after the birth, the mother will cover her offspring with leaves

The grey squirrel (*Sciurus carolinensis*) lives in a tree hole or a nest made from twigs and other debris, built in the branches of a tree and located well off the ground.

The two species of red squirrel (*Tamiasciurus* species) occur in North America. Found in areas of forest, they can also swim well on occasions. This is the eastern form (*T. hudsonicus*), distinguishable from the Douglas squirrel (*T. douglasii*) by its white underparts.

when she goes foraging for food. This helps to keep them warm, as well as concealing them from possible predators. The young squirrels leave the nest for the first time when they are about 6 weeks old.

Females may breed twice during the year in some parts of their range. In the UK, breeding may occur both during the winter and in the middle of summer. There is no strong pair bond, however, and the partners separate after mating.

There has been particular concern in the UK over the introduction (see page 24) of the larger, more assertive grey squirrel (*S. carolinensis*), which has been blamed for a decline in the red squirrel population. This is not a true reflection of the situation, however, although relatively few areas of the UK, apart from the Isle of Wight, are free from grey squirrels. It is really more a matter of habitat: red squirrels by nature prefer coniferous woodland, whereas grey squirrels are far more opportunistic, successfully invading suburban gardens, and are more commonly seen.

Feeding predominantly on Scots-pine cones, red squirrels need mature stretches of woodland, where the trees must be at least 35 years old, to guarantee them a constant food supply. Deciduous woodland is far more suitable for grey squirrels, which are not inclined to penetrate into coniferous areas. The key to ensuring the survival of the red squirrel therefore depends far more on the planting and maintenance of large-scale coniferous woods, with a minimum area of 100 hectares (250 acres), than curbing the numbers of grey squirrels.

The white area of fur surrounding the eyes of these red squirrels is especially prominent, as shown by this photograph. They are heavily hunted for their fur in various parts of their range.

Bolder markings are a feature of some species, such as the Asiatic striped palm squirrels (*Funambulus* species), whose distribution is centred on India. Three dark stripes are fairly typical in this case.

Chipmunks are ground squirrels, which live and store their food in underground burrows, although they can climb on occasions and sometimes nest off the ground. This is an Asiatic chipmunk (*Eutamias sibiricus*), which occurs on the far east of the continent.

Where they do occur together, red squirrels can be encouraged to breed by the provision of nest boxes which are too small to allow the entry of greys. Other simple steps can be taken to safeguard populations of red squirrels, by providing safe, shallow drinking containers and additional food, like maize, when their natural food is in short supply.

American red squirrels also prefer coniferous plantations but they are also found in deciduous woodland. The eastern red squirrel (*Tamiasciurus hudsonicus*) is widely distributed in Canada whereas the Douglas red squirrel (*T. douglasii*) is found in western North America.

The latter has evolved a unique way of ensuring a food supply through the winter. It feeds mainly on pine cones and, in the autumn, collects them, cutting them off the trees before they have ripened. The green cones are then carried to a spot close to a stream or equally damp locality where they can be stored. The moisture prevents the cones opening, leaving the seeds safe within. Over 150 cones may be stored at a single locality and these squirrels will return to the same place to feed, burrowing through as much as 4 m (13 ft) of snow to reach the cones during the winter.

Huge numbers of American red squirrels – up to 3 million annually – are killed in Canada for their fur. They can be prolific, with females averaging five offspring in a litter and breeding occurring twice a year.

It is not just coincidental that squirrels rank among the most colourful of rodents. They are one of the few groups to possess colour vision. The 18 species comprising the genus *Callosciurus* include some of the most attractively patterned members of the family. Their colours can include light cream, reddish brown and jet black.

Although sometimes confused with striped squirrels, chipmunks (*Tamias* and *Eutamias* species) form a separate group. Their distribution is centred on North America but one species, the Siberian chipmunk (*E. sibiricus*), is present in parts of eastern Asia.

Chipmunks tend to burrow, storing food and spending part of the winter underground when the weather is at its worst. They will also climb and are agile in the trees, running quickly from branch to branch. Active during the day, chipmunks are not particularly social by nature and may be territorial, especially close to their burrows.

Marmots (*Marmota* species) live in even harsher climates than chipmunks and their larger size may help to protect them from the cold. They are to be found in the northern USA, including Alaska, Canada and through much of northern Asia to mountainous areas of Europe, notably the Alps.

These rodents excavate a deep and extensive network of underground burrows, where they spend much of the year. Tunnels may be as long as 113 m (370 ft) and the chambers may be 7 m (23 ft) below the surface. Relatively open countryside is favoured by marmots and, although they tend to stay on the ground, they can climb and are able to swim if necessary.

Marmots feed mainly on greenstuff but also eat seeds, fruits, and even insects on occasions. They build up body stores of fat during the brief

Marmots (*Marmota* species) live in fairly inhospitable parts of North America and Eurasia, spending much of the winter hibernating in their burrows. When above ground, their fur helps them to merge into their background, as shown by this individual.

The entrance to the marmot's burrow may be well concealed. Burrows that are used for hibernation purposes may be 7 m (21 ft) below the ground.

Occupying areas of open country, prairie dogs (*Cynomys* species) can occur in huge numbers in favourable habitat but have become much scarcer in recent years, having been persecuted by people in many parts of their range.

summer period and may then hibernate underground for as long as 9 months.

Although marmots live in colonies, their sociability varies according to the species. Breeding starts almost as soon as they emerge from their winter sleep and pregnancy lasts for about 4 weeks. Some species of marmot only breed in alternate years. Colonies often consist of a male and several females with their youngsters. In remoter parts of their range, marmot populations are still healthy but, particularly in Europe, agricultural incursions have adversely affected their numbers.

Prairie dogs (*Cynomys* species) are the most highly social of all rodents, living in the grasslands of North America. They live in colonies, appropriately called 'towns'. These communities can be enormous. One such group in western Texas was reckoned to consist of 400 million black-tailed prairie dogs (*C. ludovicianus*). It spread over an area of 64,000 km², (24,700 sq. miles). Each town consists of smaller sub-groups, or 'wards', which in turn are split into a number of 'coteries', basically family units.

Their feeding habits soon have an impact on the vegetation in an area. Tall plants are destroyed, as are those which grow slowly. The clear space which is thus created makes it difficult for predators to creep up

Ground squirrels or susliks (*Spermophilus* species) have a wide distribution, both in North America and Eurasia, inhabiting open areas of countryside rather than forests. Spots rather than stripes are a common feature of the group.

undetected on the colony. If they are threatened, prairie dogs retreat underground.

They do not store food in their burrows, nor do they hibernate for any length of time during the winter months. Females breed just once a year, producing an average litter of four offspring. The young prairie dogs first emerge from their burrow when they are 5 weeks old and are weaned soon afterwards.

Prairie dogs are viewed as pests by many farmers and huge numbers have been killed by the widespread use of poisons. This has now resulted in some species, such as the Utah prairie dog (*C. parvidens*), becoming rare, while widespread habitat change, particularly increasing urbanization, is also exerting a deleterious effect on the numbers of these rodents.

Changes in agriculture are not necessarily harmful to sciuromorphs and even prairie dogs benefited for a time in some areas where cattle-grazing kept the vegetation low. Ground squirrels, or susliks (*Spermophilus* species), are another group whose numbers have grown as a result of the clearance of forest for agriculture. In Canada, these ground squirrels have become major pests and their numbers are having to be controlled by the use of poisons.

SCALY-TAILED SQUIRRELS • FAMILY ANOMALURIDAE

The unusual name of these squirrels stems from the unique structure of their tails. These are covered with hair on the upper surface, while beneath, starting at the base of the tail, are a series of overlapping scales which extend for up to one third of the length of the tail. Their purpose is to provide extra support when these squirrels are climbing around in the trees.

Seven species of scaly-tailed squirrel are known, all of which are found in parts of western and central Africa. Scaly-tailed squirrels possess a membrane down each side of the body, enabling them to fly, with the notable exceptions of the aptly named flightless scaly-tailed squirrel (*Zenkerella insignis*). The arrangement of these membranes differs from that of true flying squirrels in that they are attached not to a bone protruding from the wrist joint but to a rod of cartilage extending from the elbow. The tail scales may also serve to provide extra support on landing, in addition to the strong claws.

Very little is known about the habits of this family of squirrels, although they appear to be nocturnal, like other flying squirrels. They may live in groups of up to 100 individuals in areas where food is plentiful.

The largest is Pel's scaly-tailed squirrel (*Anomalurus pelii*), which measures 45 cm (18 in) long, with a tail of similar length. These squirrels appear to be entirely arboreal and, if placed on the ground, have great difficulty in moving, being handicapped by their wing membranes. Under these circumstances they will hop to a nearby tree and start to climb. When gliding, the larger members of the group have been

Distribution of scaly-tailed squirrels (Anomaluridae).

observed to travel distances of up to 95 m (310 ft), although generally they just move from tree to tree.

Vegetarian in their feeding habits, scaly-tailed squirrels eat plants, flowers and nuts. They roost in tree hollows and females give birth to a single youngster in these surroundings, although there is little detailed information known about their breeding habits.

SPRINGHARE • FAMILY PEDETIDAE

The only member of this family lives in southern Africa, inhabiting open and fairly arid countryside. Looking rather like a kangaroo, the springhare (*Pedetes capensis*) has long and powerful hind legs, with four toes on each of its feet, terminating in large, rather hoof-like claws. The forelegs in comparison are very short, averaging just one quarter the length of its hind legs.

As might be expected, the springhare moves mainly by hopping on its hind feet, with its long, dark-tipped tail helping to keep its balance. When sitting on its hind legs in an upright position, a springhare is approximately 30 cm (12 in) high. Its overall body coloration is sandy brown, with white underparts.

The springhare has a long head with prominent ears, reminiscent of a rabbit, and this, coupled with its hopping gait, explains how the species acquired its name. It weighs 3–4 kg (6½–9 lb).

The large eyes of the springhare indicate that it is active after dark. During the daytime, these rodents hide away from predators in burrows which they dig in dry, sandy soil. These can be quite large, extending to a depth of about 80 cm (32 in) and covering an area of up to 120 m^2 (1,200 sq. ft), with individual tunnels of up to 46 m (50 yd) long.

The springhare is well adapted to a subterranean existence. It can fold back its ears and seal its nose while digging, in order to exclude dirt. Its short forelegs, equipped with five sharp claws, are ideal for excavating underground. At intervals, the springhare will stop and turn round in its burrow, pushing the accumulated soil back with its legs and chest. The hind legs are then used to kick the soil and disperse it above ground.

Most springhare burrows are roughly circular, with many entrances: as many as eleven to a single burrow. These help the springhare to escape if predators, such as mongooses, enter the network of tunnels. The springhare itself appears to move into different parts of the burrow and will sometimes seal an entrance behind it.

Each system of burrows is occupied by a single springhare. Females give birth to just one youngster, with breeding taking place throughout the year. At any time, about three-quarters of the female springhare population is likely to be pregnant. They can have three litters per year, the young being born after a gestation period of about 80 days.

At this stage, the baby springhare is well developed, with its eyes open, and weighs about 300 g (10½ oz). It will be suckled and remain below ground until it is nearly 7 weeks old, when it will be nearly fully grown.

Also sometimes called the springhaas, the springhare (*Pedetes capensis*) holds its tail upwards when it jumps. It digs its burrows in sandy soil, with burrowing activity increasing after the rainy period, when the sand is firmer.

Young springhares must then find a safe burrow for themselves. In areas of suitable habitat, where food is freely available, there may be over 10 per hectare (2½ acres).

Springhares forage on all fours, digging up roots and other vegetation. They feed under cover of darkness, rarely venturing far from their burrows when the moon is full. They tend to associate more above ground where their keen senses can alert the group to the approach of a potential predator.

In spite of their relatively large size, these rodents are vulnerable to many predatory species, ranging from owls to lions. They are also regularly hunted by Bushmen, who greatly value these rodents as a source of food and other supplies, such as fur for clothing.

In order to escape being caught, the springhare can jump up to 4 m (13 ft) at a single bound and, at close quarters, will both bite and kick ferociously with its sharp-clawed hind feet.

In some parts of its range, the springhare causes significant damage to crops but, even where they are hunted heavily, these unusual rodents still appear to be quite numerous.

BEAVERS • FAMILY CASTORIDAE

After the capybara (*Hydrochaeris hydrochaeris*), the beavers are the second largest of all the world's living rodents. They can weigh as much as 30 kg (66 lb) and stand 60 cm (24 in) tall at the shoulder. Two species survive today, the North American beaver (*Castor canadensis*) being more widespread that its Eurasian relative, *C. fiber*.

The waterproofing provided by the beaver's coat is clearly illustrated by this North American beaver (*Castor canadensis*), which has just left the water. The coat looks smooth and shiny as a result.

Beavers have a unique appearance, being thickset with a highly unusual flattened tail. They are superbly equipped for their predominantly aquatic existence. Apart from their horizontal scaly tail, which acts like a flipper, providing power and a means of steering underwater, they also have powerful webbed hind feet. Their small eyes are protected by a translucent membrane when they are beneath the water's surface. Their nostrils can be closed to prevent water entering, and so can their small ears.

Beavers live close to water and are one of the few mammals, apart from human beings, which actively seek to modify the landscape to suit their needs. In this respect, the beaver's powerful incisors and massive skull, providing support for the muscles, are essential attributes.

Seeking out streams with trees such as willow, aspen, alder and birch growing in the vicinity, beavers use their teeth to fell suitable saplings and use the cut trunks to dam the stream or river at a suitable point. This holds back the water and provides a deeper and more tranquil area where the beavers can construct a lodge.

The dam must be built carefully, on a suitable bed of mud and stones, which will serve as anchorage points for the saplings. Other vegetation and mud are used to hold the dam together and repairs are carried out on a regular basis.

Once a pair of beavers has constructed a dam, subsequent generations usually add to it so that, ultimately, a large structure will be built up, with the dammed water extending over a wide area.

The largest beaver dam ever recorded, across the Jefferson River, in Montana, USA, was 700 m (2,300 ft) long and wide enough to be crossed on horseback. Typical dams are likely to be 23 m (75 ft) long and hundreds of saplings may be felled to build them.

Beavers tend to cut trees more in the autumn and winter, before the sap starts to flow, probably because it is easier then. As the water dams up, a pair will start to construct their lodge in a similar fashion, using logs and vegetation. Beavers may even build special canals to help them float trees to the site of the lodge or dam.

Just as the dam will be higher than the water level, so the lodge will grow in height, sometimes standing 2 m (6½ ft) above the surrounding water, with a diameter of more than 12 m (13 yd). Entrances concealed beneath the water lead into the lodge, where there is a single large chamber, about 2 m (6½ ft) wide, protected by thick walls. This is where the beavers live and where the female gives birth to her kits in the early summer, after a pregnancy lasting about 14 weeks.

Young beavers are born with their eyes open and have a covering of fur. They are fully weaned at 3 months old and subsequently stay with their parents for 2 years or more. A beaver colony therefore consists of a family group, the young beavers only moving elsewhere during their second year, when they are ready to start breeding themselves. In most cases they travel no further than 20km (12½ miles).

Beavers feed on bark, shoots and leaves of trees, as well as on aquatic plants. Underwater, close to the lodge, they store wood for food over the winter period. Here it is within easy reach, even if the water surface is frozen, and it can be retrieved safely without the beavers having to venture onto the ice.

Various animals, such as wolves, prey on beavers but hunting of beavers for their dense, water-repellent fur has led to the deaths of millions of these rodents, particularly in North America, where they are most common today. Beavers are clumsy on land and, if danger threatens, they scurry back to water. The tail can be used to slap the surface as a warning signal and, if necessary, they can remain submerged for 15 minutes or more.

Scent-marking is particularly important in this group of rodents. It is used to emphasize territorial boundaries, especially in spring. The special castor glands and anal glands are employed for this purpose. Secretions from these glands are passed into the cloaca – the beaver is the only mammal to have such a structure – from where they are deposited on prominent rocks and other sites.

POCKET GOPHERS • FAMILY GEOMYIDAE

The description 'pocket' refers to the fur-lined cheek pouches present in this family of rodents. There are 34 different species in this family, whose distribution ranges from western Canada south into parts of Mexico and Panama. The ranges of the pocket gophers are well defined and there is very little overlap between them. Even so, taxonomists have encountered difficulty in separating them, with in excess of 185 subspecies of the valley pocket gopher (*Thomomys bottae*) having been identified. Pocket gophers inhabit areas of open countryside, where they can dig their burrows without difficulty.

The thick-set, stocky bodies of the pocket gophers are ideally suited for a subterranean life style. One of their most striking features is the prominent incisor teeth, which protrude from the mouth even when it is closed. This enables them to use their teeth for digging purposes, with no risk of swallowing any soil. They rely mainly on their front feet to excavate their tunnels, using their teeth to cut through obstructions, such as roots, which may block their path. The tremendous wear on these claws and teeth is reflected in their correspondingly high rate of growth, which can be up to 1 mm (almost ¹⁄₁₆ in) daily in the case of the valley pocket gopher.

In common with other burrowing rodents, the eyes of pocket gophers are small and can be kept tightly closed to prevent specks of dust entering the eyes during the digging process. Similarly, their ears are small and the ear canal can be shut off when required.

As soil is excavated, so the rodent pushes the earth underneath its body, finally turning around to move it, with its feet, out of the burrow. The depth of the tunnels depends on the condition of the soil; they are closer to the surface where the ground is hard. The nesting chamber is situated at the deepest point, while the outer entry point to the burrow is kept sealed at ground level.

Pocket gophers feed mainly on plant matter, with *Geomys* species rarely venturing above ground, feeding instead on roots and tubers. Others of the genus *Thomomys* leave their burrows after dark and forage for food on the surface. Here they also collect grass, which is used to line the nesting chamber.

These rodents can cause serious damage in some agricultural areas and, in Mexico, farmers rely on the services of professional trappers, known as *tuceros*, to control the numbers of pocket gophers on their land.

Pocket gophers tend to be territorial by nature and will fight to the death if housed in close proximity to each other. Only during the breeding season do they become social, with males often establishing harems of several females during this period.

The litter size varies considerably, depending on the habitat, and between three and six youngsters are usually produced. They develop slowly, with their eyes and ears not opening until they are nearly 4 weeks old. Young pocket gophers finally leave the nest, along with the male, when they are about 2 months old.

The bare tail of these rodents seems to be highly significant in the regulation of their body temperature. Although they have thick fur their tail allows them to lose body heat and, moreover, is longer in those individuals living in warmer parts of their range.

In terms of size, they vary from 12 to 30 cm (5 to 12 in) long, depending partly on the species concerned, although male pocket gophers can be up to twice as large as females. There is variation within populations as well, with those found at higher altitudes tending to be larger than their lowland relatives.

POCKET MICE • FAMILY HETEROMYIDAE

Like pocket gophers, these mice also have external cheek pouches lined with fur. These are used for carrying food and can be turned inside out for cleaning. Five different genera make up this family and there is considerable variation in their appearance.

The kangaroo mice (*Microdipodops* species) which are found in the western USA as far east as Utah, have short front legs and powerful elongated hindquarters. They move entirely on their hind legs, hopping like kangaroos. Significant anatomical changes can be seen in the skulls of both species of kangaroo mouse. The tympanic bullae, responsible for hearing, are greatly enlarged, while the skull itself is very thin. The presence of stiff hairs along the edges of the hind feet allows these rodents to jump at speed across sandy areas without sinking into the sand.

The tail, which is usually kept off the ground, is used for balancing and also has a fat store as its base. For part of the year, these small desert-dwelling rodents may hibernate in their shallow burrows, which are often concealed by the roots of shrubs.

Kangaroo mice are strictly nocturnal and dislike light. They drink

Distribution of pocket mice (Heteromyidae).

very little water, deriving their fluid requirement from their metabolism and any water present in their food. They feed mainly on seeds but also eat insects on occasions.

There are 22 species of kangaroo rat (*Dipodomys* species), which range over a much wider area of North America, from southwestern Canada to central Mexico, than kangaroo mice. Larger in size, with a maximum body length of 20 cm (8 in), and a tail of similar length, they appear to have a similar life style and also move by hopping on their well-developed hind legs. They can cover 2 m (6½ ft) at a single bound.

Kangaroo rats are also nocturnal by nature and have proved to be highly territorial, with adults having their own burrows. Their kidneys are said to be four times more effective at conserving water than human kidneys, enabling them to produce a very concentrated urine.

The breeding season varies depending on the availability of food. Under favourable conditions, kangaroo rats can breed throughout the year. Pregnancy lasts about 4 weeks and there are three youngsters in the average litter. They will leave the nest when they are just over 1 month old.

Unfortunately, a number of kangaroo rats have very limited areas of distribution and are sensitive to habitat changes. This has led to a decline in the populations of several species, particularly those found in California.

Spiny pocket mice (*Liomys* species) range throughout Central America, extending from southern Texas and Mexico to Panama. They are paler in colour and slightly larger in size than forest spiny pocket mice (*Heteromys* species), which, as their name suggests, tend to occur

The short front legs, powerful hind limbs and long tail characterize the kangaroo rats (*Dipodomys* species), whose distribution is centred on arid parts of the western USA.

predominantly in areas of woodland. The bristly coat is also more highly developed in the spiny pocket mice. Both genera of spiny pocket mice have a spoon-like claw on the second digit of each hind foot, which may be used for digging or grooming, but this is less conspicuous in the forest spiny pocket mice.

Members of both genera live in underground burrows and emerge at night in search of seeds, nuts and similar foods, which are carried back to the burrow in their cheek pouches. There is some evidence to suggest that *Liomys* species are far less social by nature than *Heteromys* species, which may live communally in their burrows. Representatives of *Heteromys* are also found in northern South America, extending southwards from Mexico.

Pocket mice of the genus *Perognathus* are confined entirely to North America. The hind limbs are only slightly longer than the forelimbs and these mice have a quadrupedal gait, with their hind legs simply providing support as they dig in the sandy soil where vegetation is sparse.

Nocturnal by nature, and remaining hidden underground during the day, when the ground can become very hot, pocket mice seal the entrances to their burrows to keep themselves cool. They may remain underground for months at a time, relying on the food stores carried to the burrows in their cheek pouches.

These mice are careful to conceal the entrances to their underground homes beneath boulders or plants and they scatter the excavated soil rather than letting it accumulate in one place.

Breeding success within a population depends to a large extent on the availability of food. In good years, females may have three litters, each comprising as many as seven youngsters but, when food is in short supply, only one third of the females is likely to give birth. A study of *P. parvus* revealed that, in the wild, most youngsters survive for up to a year but barely three out of every 100 survive for as long as 4 years.

Chapter 7
Myomorpha: Mouse-like Rodents

This group is not only the largest of all three rodent suborders, it also embraces over one quarter of all the mammals found on the planet today. As might be expected, representatives of this group can be found worldwide, on all continents. The house mouse (*Mus musculus*) has even invaded Antarctica successfully, thanks to its close association with people (see page 11).

All members of this group are quite small in size. They range from the harvest mouse (*Micromys minutus*) from Europe and Asia, which weighs just 6 g (¼ oz) when adult, to the Cuming's slender-tailed cloud rat (*Phloeomys cumingi*) from the Philippines. This species can grow to 83 cm (33 in) long, including its tail of 35 cm (14 in), and may weigh 2 kg (4½ lb).

Myomorphs have successfully colonized almost every type of ecosystem, from the Arctic to temperate woodlands, and deserts to tropical rainforest. This suborder is split into five families, of which the largest by far is the Muridae, comprising rats, mice and other similar rodents, such as hamsters, gerbils and mole rats. Not all of these groups are generalist and adaptable. Some, such as the Australian water rats (subfamily Hydromyinae) are highly specialized in their habits.

It can be misleading to rely simply on common names when ascribing rodents to a particular suborder. There are many so-called 'rats', such as the caviomorph spiny rats, which, in fact, belong to a different suborder.

RATS AND MICE • FAMILY MURIDAE

New World rats and mice • Subfamily Hesperomyinae
Members of this group first developed in North America, invading South America when the link between these two continents was forged for the second time about 5 million years ago. One of the most widely distributed myomorphs in this region today are the rice rats (*Oryzomys* species). They range from the south-eastern USA, through Central America, to Tierra del Fuego, at the very tip of South America. Rice rats have also crossed to the Galapagos Islands, off the western coast of Ecuador, where they have become established.

The spread of rice rats into South America reveals the way in which the myomorphs especially can diverge and multiply rapidly under favourable conditions, with species establishing individual niches. In Venezuela, for example, there are seven different species of rice rat, occurring in different types of habitat. While *Oryzomys capito* is found

Distribution of mouse-like rodents (Myomorpha).

near homes, often close to water, *O. bicolor* is more arboreal in its habitats and is sometimes found in drier parts of the country.

New species are still being discovered, such as the silver rice rat (*O. argentatus*), which was obtained on Cudjoe Key, off southern Florida, in 1973. It appears to have a very localized distribution and faces an uncertain future since the recent development of its habitat.

Diversity brings casualties and a number of *Oryzomys* rice rats in the Caribbean and Galapagos have already become extinct. Closely related were the West Indian giant rice rats (*Megalomys* species) which are also now extinct (see page 96). These species probably evolved from *Oryzomys* stock. The closest living link is believed to be *O. hammondi*, which now inhabits the Ecuadorian Andes and was probably the ancestral form from which many of today's species are derived.

A surviving genus which is clearly related to *Oryzomys* is *Neacomys*. Three species, ranging from Panama into northern South America, are recognized. They are collectively known as spiny rice rats and the only distinction between them and *Oryzomys* is the texture of their coat. This comprises a combination of bristles, which predominate on the back but become less conspicuous on the flanks and are completely absent from the belly, and softer, thinner hairs.

Found further south, in southeast Brazil and neighbouring parts of Argentina, is another species, *Abrawayaomys rushcii*, first discovered in 1979, which appears to link both *Neacomys* and *Oryzomys*. It has a spiny coat, with an arrangement of spines similar to that in *Neacomys*, but is much larger. An obvious relationship with *Oryzomys* is indicated by the pattern of its molar teeth.

While rice rats are often found close to water, some members of the

Muridae have evolved further towards an aquatic life style. The Neotropical water rats (*Nectomys* species) are to be found in northern and central South America. Their glossy coat, with a dense undercoat, gives good waterproofing when they swim. These rats inhabit areas of forest close to rivers and streams, and are powerful swimmers. Their hind feet are large, with webbing between the toes. They also have bristly hair around the feet as well as on the lower surface of the tail. Neotropical water rats swim regularly in search of food and eat snails and arthropods, as well as fruit, seeds and similar items.

Rather similar in their habits are the web-footed rats (*Holochilus* species), which are more widely distributed through South America, extending as far south as Uruguay and northeastern Argentina. Their nests, built of aquatic vegetation, may be located up to 3 m (10 ft) off the ground, whereas Neotropical water rats conceal their nests on the ground, burrowing under old logs and vegetation.

The nest of the web-footed rat may be up to 40 cm (16 in) in diameter and may weigh as much as 100 g (3½ oz). The rats live in the top portion while the lower part is lined with vegetation. These nests may be built in close proximity to each other and the rats, if frightened, will scamper out of them into the water and swim off.

Web-footed rats are mostly nocturnal, feeding on aquatic plants and snails, as well as seeds gathered on land. Their numbers can increase dramatically when bamboo in any area flowers and seeds and, once this glut of food is utilized, they may switch their attention to crops.

Several genera of water rats in South America have become more aggressive hunters. These are the fish-eating rats, of which *Ichthyomys* is the most highly specialized. Large, partially webbed hind feet, complete with bristles around the edges, and a tail with bristles on its lower surface help these rats to swim effectively. Their body is also more streamlined, with the ears and eyes being relatively small. The whiskers take over some of the sensory functions, being both long and quite rigid.

Fish and other aquatic creatures are seized in the jaws of these predatory rats and impaled on their incisor teeth. The outer edge of each tooth is formed into a point for this purpose, to give an improved grip. These rats are capable of catching and killing fish at least 15 cm (6 in) long, which is almost equivalent to their own body size.

During the day, fish-eating rats are reputed to conceal themselves under rocks close to the streams and swamps where they hunt. Very little seems to be known about their breeding habits.

An even better swimmer than *Ichthyomys* species can be found in upland areas in northern Ecuador. This aquatic rat, known as *Anotomys leander*, has tiny ear flaps and extra hairs on its feet to allow it to swim more powerfully. It also has a higher diploid chromosome count than any other mammal, possessing 92 chromosomes.

Another aquatic rat, with a limited distribution in the Andes of northern Ecuador, is *Neusticomys monticolus*. It is most numerous in areas of relatively fast-flowing water, although it is less well adapted for an aquatic life than other water rats. It is easily distinguished by the first digit on each of its hind feet, which is much smaller than that of other water rats.

It is not just rats which have adapted to an aquatic existence. Central American water mice (*Rheomys* species) can be found from southern Mexico southwards into Colombia and neighbouring parts of western Venezuela. The most obvious feature of these mice is their streamlined shape and large, webbed hind feet. The forelimbs are relatively small in contrast. Extra propulsive power is provided by the tail. The ears and eyes are small and the nose can be closed by means of flaps when the mouse is in the water. Central American water mice live near fast-flowing streams in rainforest areas. Here they hunt for aquatic invertebrates, including snails, and may catch small fish.

Other New World murids have also evolved to suit arboreal existence. A typical example of this group is the big-eared climbing rat (*Ototylomys phyllotis*), which lives in parts of Central America. It typically inhabits areas of forest where there are rocky outcrops and, although it appears to spend much of its time off the ground, it will also live among the rocks.

This species is so-called because of its large ears, which are basically free from hair. Keen hearing may help these rats to avoid danger by scurrying to a suitable tree hole or other cover. Other climbing rats belonging to the genus *Tylomys* are more widely distributed in Central America. They have been observed in palm trees up to 9 m (30 ft) off the ground.

Some mice have also become adapted to seek food and shelter off the ground and the Brazilian arboreal mouse (*Rhagomys rufescens*) is a typical example. Occurring in eastern Brazil, these mice are lightly coloured, with an orange-rufous coat that becomes slightly paler on the underparts. Their small ears are heavily furred. Most significantly, the innermost first digit on the hind feet does not normally terminate in a claw, while the outer toes on each foot are long, helping the mouse to retain a grip as it walks on a branch.

Not all mice climb in tall trees, however, and American harvest mice (*Reithrodontomys* species) may simply nest off the ground in grasses or low shrubs. They are light, weighing as little as 6 g (¼ oz), and agile, and occupy an evolutionary niche similar to that of the Old World harvest mouse (*Micromys minutus*).

Their nests, which measure up to 17.5 cm (7 in) in diameter, are built of grass and they feed largely on seeds. These mice, which range from the southern USA through Central America, may be prolific breeders. In captivity at least, females have been known to produce as many as 68 offspring in the course of a year, although very few live for this long in the wild.

One of the most unusual harvest mice is the endangered saltmarsh species (*R. raviventris*), which lives in the salty marshland of San Francisco Bay. It can survive here by drinking salt water but, unfortunately, habitat changes are threatening the future of this rodent.

The grasslands of the New World have also been successfully colonized by various different mice. Grasshopper mice (*Onychomys* species), which are largely confined to North America, are found in such areas. Although they will eat seeds, these mice are more predatory by nature, preferring to feed on various invertebrates, such as beetles and grasshoppers. One of the three recognized species, *O. torridus*, actively hunts scorpions and it is not unknown for these mice to prey on smaller rodents as well.

Like bigger predatory mammals, grasshopper mice have quite large territories and do not occur at high densities. They have a shrill call, which can be heard up to 100 m (110 yd) away. This is often uttered while the mouse is standing on its hind legs, just before it makes a kill.

South American field mice (*Akodon* species) do not appear to be as carnivorous as grasshopper mice but invertebrates still form a significant part of their diet. In spite of their name, they are widely distributed across the continent and are sometimes encountered in areas of forest.

Wood rats (*Neotoma* species) are found in North and Central America, but again their common name is misleading, as they occur in areas of desert through to upland rocky terrain where tree cover is sparse. Nocturnal by nature, they have furry tails which give them a rather squirrel-like appearance, an impression reinforced by the fact that they can also climb if necessary.

They build remarkable nests, which are maintained and enlarged by subsequent generations. These nests take about a week to construct and may reach a height of more than 2 m (6½ ft). Those of *N. lepida*, for example, are built off the ground at heights up to 7.5 m (25 ft) and may even overhang a stretch of water. Given a choice of materials, the rats appear to be highly selective about what they use. They prefer bright objects and will discard a dull piece which they have been carrying if, for example, they come across a shiny bottle top. Because of this strange habit, these rodents are sometimes referred to as 'pack rats'.

Where natural cover is available, it is used to provide added protection for the nest. For example, the nest may be constructed around a spiny cactus, so that the rats can move in and out with impunity, while the spines provide a painful deterrent to any predator. The nest contains separate chambers for sleeping and food storage.

Some of the most adaptable and widely distributed of the New World rodents are the deer mice (*Peromyscus* species), which range from Canada to Mexico. They are sometimes called white-footed mice, because of their white underparts. In all species, the ears are relatively large but those which inhabit open areas of countryside are paler in coloration than those found in woodland. Under favourable conditions, deer mice will breed at any stage during the year. They use their front feet to bang on the ground if excited.

The leaf-eared mice (*Phyllotis* species) of South America have even larger ears than deer mice, which they otherwise resemble, and seem quite adaptable in their habits. These mice often hide in spaces between rocks and are found in fairly arid surroundings.

Other mice are also found at high altitudes in the Andean region, including the altiplano chinchilla mouse (*Chinchillula sahamae*), which, like the chinchillas (*Chinchilla* species), is hunted for its dense, warm fur. It takes the skins of over 150 of these mice to make a garment and heavy trapping for this purpose has resulted in localized declines in their numbers.

Chinchilla mice have a fairly rigorously defined breeding season in the southern spring, when greenfood is most readily available. They will gorge themselves on this food, presumably eating seeds and other vegetable matter as well.

Other chinchilla mice, belonging to the genus *Euneomys*, are found in the Patagonian region of South America. Their ears are noticeably smaller and they are more active burrowers than their altiplano relative.

Adaptations for underground life can be seen in a number of other New World genera, particularly those found in South America. Burrowing mice (*Oxymyeterus* species) show the typical features of this group: a long, narrow snout and well-developed claws on their forefeet. They hunt mainly for insects below the surface of the soil.

Even more marked in terms of its appearance as a burrowing rodent is the Brazilian shrew-mouse (*Blarinomys breviceps*). Its ears are greatly reduced in size and hidden by fur, and its eyes are also tiny. Living in rainforest areas, these rodents tunnel into the ground, with the burrow tending to level off below ground.

Mole mice (*Notiomys* species) construct an even more elaborate network of tunnels. They have particularly well-developed claws on their front legs and, as in the Brazilian shrew-mouse, their tail is short. They rarely emerge above the surface.

Voles and lemmings • Subfamily Microtinae
Members of this subfamily are confined to the northern hemisphere, extending from the Arctic to Central America and the Himalayas in Asia, with one species occurring in Africa. The majority of the species in this group are burrowers by nature and are relatively small. They average about 13 cm (5 in) in length, including their short tail, and weigh up to 20 g (¾ oz). The notable exception is the muskrat (*Ondatra zibethicus*), which lives in parts of North America, and can weigh at least 1.8 kg (4 lb).

The muskrat is so-called because of the musky odour liberated from glands in the perineal region. The secretions are used in scent-marking its territory. It is an aquatic species, using its webbed feet for swimming, and spends long periods in the water. Although it normally surfaces to breathe every few minutes, the muskrat can remain submerged for as long as 17 minutes at a time.

The muskrat may burrow into a suitable bank below the waterline, creating a concealed tunnel that can extend back for 10 m (11 yd). Alternatively it may build a house of vegetation, piling this up and plastering the outside with mud. It feeds largely on aquatic plants, although it also eats crabs, fish and similar items.

Pairs may produce two litters each year and the youngsters remain with their parents until they are mature, which may take as long as a year in some northerly parts of their range.

The Florida water rat (*Neofiber alleni*) is sometimes called the round-tailed muskrat although, apart from the shape of its tail, it is significantly smaller than the muskrat, weighing a maximum of 350 g (12½ oz). Its feet show little sign of webbing and it spends more time out of the water.

The Florida water rat builds a large house, using grass and other vegetation, with a central nesting chamber and underground exits. It feeds entirely on plants. Females can be prolific and may rear five litters in a year, producing just over two young in each on average.

Although they are not hunted, Florida water rats have declined in numbers in some areas because of changes in the water table, which have allowed salty water to enter their marshland habitat.

Other, smaller members of this subfamily have also opted for an aquatic existence, such as the water vole (*Arvicola terrestris*), which is widely distributed in western Europe. It is the largest of the Old World microtines, similar in size to a rat, with a relatively short tail, about half the size of its body. A related species, *A. richardsoni*, is found in North America.

These water voles excavate their burrows in suitable areas adjoining stretches of water, creating an entry point to the tunnel below the water surface. Vegetarian in their feeding habits, they eat a variety of aquatic plants. They can grind up tough vegetation quite easily using their powerful molar teeth, which continue to grow throughout their lives.

Some populations of water voles are found well away from an aquatic environment, however, particularly in upland areas of Germany and Switzerland. These voles have adapted to a fossorial life style, burrowing

The muskrat (*Ondatra zibethicus*) builds retreats made of aquatic vegetation plastered with mud. Once the young muskrats are independent, they help the adult pair to maintain this home.

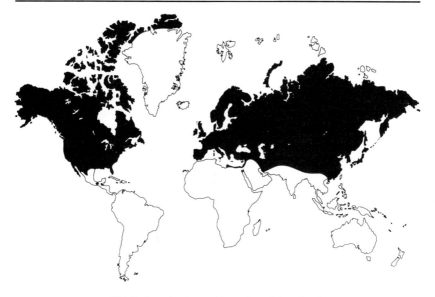

Distribution of voles and lemmings (Microtinae).

underground. They still closely resemble their aquatic relatives, although they may be somewhat lighter in coloration and the upper incisors are positioned at a more horizontal angle, presumably to assist with the more demanding excavation of their burrows in such surroundings. Their underground tunnels regularly exceed 34 m (37½ yd) in length and food storage chambers are incorporated in them.

Water voles have prominent scent glands on their flanks, which they use to alert other voles to their territorial boundaries. Males may fight quite ferociously if challenged. Their glands enlarge at the start of the breeding season and the secretions which they produce are smeared on their feet and then rubbed into the ground.

The reproductive rate of these voles is high; females have five litters and produce as many as 30 offspring during spring and summer. Their life expectancy is correspondingly short, not exceeding 6 months on average. Young water voles may be mature when just 5 weeks old and, if the weather is favourable, breeding may continue through the winter months.

In parts of Asia, various voles are found in alpine meadows. They include members of the genus *Alticola*, which can occur at altitudes approaching 6,000 m (20,000 ft). These voles store food in their burrows to last them over the winter period, having first cut and dried the herbage. In Pakistan, two similar species, belonging to the genus *Hyperacrius*, can be found: the Punjab vole (*H. wynnei*) is fossorial in its habits while the Kashmir vole (*H. fertilis*) is more likely to be seen above ground.

Further to the west, another upland species, Martino's snow vole

(*Dinaromys vogdanovi*) is found in the mountainous areas of the former Yugoslavia. Breeding appears to be restricted to the warmer summer period and only one litter of two or three youngsters is produced.

Meadow voles (*Microtus* species) are the most widely distributed genus. They have small, inconspicuous ears and broad heads, and are often greyish brown in colour. These voles are found in many different habitats, ranging from tundra to coniferous forests. They do not hibernate during the winter, although they often store food in their burrows. All are vegetarian in their feeding habits.

Their populations often show a cyclical pattern, with aggression building up as the number of voles in an area rises. The life span of these voles is short but they have a correspondingly high reproductive rate and display a rapid onset of maturity.

Females have litters comprising a dozen or more offspring, which, in turn, may breed when less than 4 weeks old. In the case of the meadow vole (*M. pennsylvanicus*), the average life expectancy may be less than a month in some areas. Huge numbers fall victim to birds of prey, foxes and other predators.

Red-backed voles (*Clethrionomys* species) form another genus with representatives in both North America and Eurasia. They feed on insects as well as vegetation, and are well equipped to eat grasses, possessing molar teeth which may continue growing for some time before developing roots. Indeed, it is possible to estimate the age of red-backed voles from the amount of wear apparent on their molars.

Studies of their breeding habits have revealed that red-backed voles born in spring mature more quickly and have a longer life expectancy than those born during the summer. They usually nest underground but may sometimes choose a suitable site under a fallen tree or elsewhere on the surface.

The mole-voles, or mole-lemmings (*Ellobius* species), which live in western Asia, are well equipped for an underground existence. Tiny ears and eyes are present on their powerfully rounded head, with large incisor teeth in both jaws. Those in the upper jaw protrude forwards and in a downward direction to enable them to excavate their burrows. These rodents feed on tubers and bulbs in the soil, rather than seeking food above ground. Female mole-voles are quite prolific, producing as many as seven litters annually, each comprising up to five youngsters. Their soft, velvety fur means that they are hunted for their skins in some parts of their range.

In contrast, the long-clawed mole-vole (*Prometheomys schaposchnikowi*) relies on its powerful forefeet to excavate its burrows. The claw on the central digit of each foot measures approximately 5 mm (¼ in) long. It feeds on plants as well as roots. At the centre of its network of burrows there is a nesting chamber which is lined with grass. Females have just two litters on average, each comprising three offspring.

The three species of so-called 'true' lemmings (*Lemmus*) are found from Scandinavia eastwards through Siberia, and in the far north of North America to the centre of British Columbia.

Microtus voles are small but they are active by nature and do not hibernate, even in northern parts of their range where the winters can be very harsh.

These lemmings are protected from the cold of their native region by dense fur, which in the Norwegian lemming (*L. lemmus*) is brightly coloured. This lemming feeds on mosses and grasses, foraging both day and night. (It is permanently light in the summer at this latitude.)

During the summer these lemmings live in underground burrows, with a nesting chamber at the end of the tunnel. At the approach of winter, the claws on the forefeet become significantly larger and are used for digging pathways beneath the snow.

Lemmings are vegetarian in their feeding habits and, on occasions, their numbers can build up and exhaust the available food supply. This causes mass migrations (see page 79), after which the population plummets; this is part of a regular cycle shown by these rodents.

Other lemmings also undergo marked variations in numbers over successive years. Steppe lemmings (*Lagurus* species), for example, show similar fluctuations, while further north, in even harsher terrain, collared lemmings (*Dicrostonyx* species) can reach densities of 400 per hectare (2½ acres) in peak years.

Their dark greyish coats are replaced by a pure white coat at the onset of winter. This is an unusual phenomenon in rodents and provides these lemmings with excellent camouflage in winter, when the ground is covered with snow. Their third and fourth claws also expand in width so that the lemming can dig more effectively in the frozen landscape. During the winter these lemmings are forced to subsist on a sparse diet of bark and twigs, until new plant growth begins once again in the spring.

A female will breed regularly throughout the summer months and will be exceedingly aggressive in defence of newly born young. At this stage, it is not unknown for them to kill male suitors, who may threaten their offspring. The young lemmings develop quickly and leave their mother's burrow when they are just over 2 weeks old.

Old World rats and mice • Subfamily Murinae

This is the largest of the myomorph subfamilies, encompassing about 408 different species. Some, such as the house mouse (*Mus musculus*) and the black rat (*Rattus rattus*) have not only spread across this entire region but have also managed to colonize the New World, thanks to inadvertent human assistance. These can be considered as 'generalist' species, able to adapt to a wise range of environments.

Others, such as the Old World harvest mouse (*Micromys minutus*), have much more specialized habits, in spite of having a broad distribution across both Europe and Asia. One of the smallest of all rodents, these mice are found in areas of tall vegetation, such as cornfields, reedbeds and ricefields.

They are well adapted for scampering up the stems of this type of vegetation, having broad feet, which are ideal for grasping, and a prehensile tail, which provides support and acts as an extra limb. Unfortunately, modern agricultural practices have meant that these mice are now much scarcer, because their nests are destroyed during the harvesting process.

The nest is hung about 90 cm (36 in) or more off the ground and is made up of three distinct layers. The inside is lined with shredded leaves and the whole structure can be built in as little as 2 days.

The harvest mouse is an example of how even rapidly reproducing species can suffer in the face of environmental change. Few of these mice live for as long as 6 months in the wild, so breeding becomes a priority. Young harvest mice are mature at about 5 weeks old, and females produce three to eight offspring per litter.

These are not the only mice which use their tails to maintain their balance. The Asian long-tailed climbing mice (*Vandeleuria* species) actually spend most of their time in trees and are far less agile on the ground.

Significantly larger are the Asiatic climbing rats (*Hapalomys* species), whose tails again exceed the length of their bodies. Their hind feet are well suited to provide them with support in bamboo and other tall vegetation. The first digit on each foot lacks a claw and can be used for gripping in opposition to the other toes. The large toe pads also help and the three long central toes provide extra anchorage.

Tree rats (*Conilurus* species) are also found in parts of Papua New Guinea and Australia. They are nocturnal by nature, hiding in tree holes, where they also nest. In parts of Northern Territory and on Melville Island, just off the Australian coast, *Conilurus penicillatus* descends to the ground and scavenges along the beach.

Even more remarkable are the Australian stick-nest rats (*Leporillus* species), which are found in southern parts of the continent. Social by nature, these rats pile up sticks to form their nests. These structures can be 1.5 m (5 ft) high and, in coastal districts, even seaweed may be used in their construction.

The house mouse (*Mus musculus*) is one of the most opportunistic of all rodents. Its small size and nocturnal nature mean that its presence can be easily overlooked, and so it has spread around the world, often unnoticed in ships' cargoes.

The small size and light weight of the harvest mouse (*Micromys minutus*) help to ensure that it is an agile climber. It weaves quite a solid nest, about the size of a tennis ball, in living grass. Their small size is obvious here, set against the human hand.

Changes in agriculture have meant that, in much of Europe, harvest mice are now much scarcer than was formerly the case, with their nests being easily destroyed. Mechanization has taken a toll on their numbers.

Some of the most bizarre and unusual members of this subfamily are to be found on a number of the Asiatic islands. They include the monkey-footed rat (*Pithecheir melanurus*), whose distribution extends from Malaysia to the island of Java and Sumatra. It has a long, soft, reddish coat, which is paler on the underparts. Its feet are modified for an arboreal existence and resemble those of primates, with each big toe having a thumb-like appearance for grasping purposes. These rats move slowly from branch to branch, using their tail to help them grip on occasions. Although these rodents do build nests, female monkey-footed rats are reputed to carry their two young with them.

The bushy-tailed cloud rat (*Crateromys schadenbergi*), found on the Philippino island of Luzon, is another arboreal species. It has a very bushy tail, as its name suggests, which is more like that of a squirrel than a myomorph. The coat is also woolly, with the result that local people hunt them for their fur. It can have a very shrill call and is not active at night.

Other, slender-tailed cloud rats (*Phloeomys* species) also occur on this island and can easily be distinguished by their tail, which, although covered in hair, is certainly not bushy.

Some of the largest rats in the world are to be found on the islands surrounding Asia, including the Flores giant rat (*Papagomys armandvillei*), which occurs on Flores, in Indonesia. This ground-dwelling species can grow to 45 cm (18 in) long, with a tail measuring another 37 cm (15 in). It has a typical rat-like appearance and is predominantly dark brown.

Other types of giant rat used to live on this island as recently as 3,500 years ago. One species, *Komodomys rintjanus*, which is only about half the size of the Flores giant rat, still survives today on smaller islands of Rintja and Padar nearby.

The island of New Guinea is also the home of two species of large rodent, the biggest being the giant tree rat (*Mallomys rothschildi*). This rodent is widely distributed in areas of montane forest. It can grow to 44 cm (17 in) long, with a tail of similar length, and may weigh as much as 2 kg (4½ lb). Living mainly in the trees, these rats feed on vegetation. Their large size means they are hunted for food and their prominent incisor teeth are especially valued for engraving.

In spite of its scientific name, the white-eared giant rat (*Hyomys goliath*) is actually smaller than the giant tree rat. It has a maximum body length of about 39 cm (15 in) and a tail of similar length. Found in New Guinea, its powerful skull and molar teeth enable it to feed on bamboo shoots in the forests where it lives. Females reputedly produce only a single youngster, but little is known about their breeding habits.

By far the largest group of rats are those forming the genus *Rattus*. These are common across Asia, where many species occur on islands. Over 15 species have been identified on Sulawesi alone and a dozen from Thailand. These rats can be broadly divided into two categories: those species which are truly wild, living in forested areas away from habitation, and those which are found alongside settlements and in agricultural areas.

Many of the wild rats have a localized distribution, which helps to explain why there are so many species on the island of Sulawesi. Their reproductive rate is generally lower than that of either the black rat (*R.*

The raised, thick hair of the spiny mice (*Acomys* species) can be clearly seen in this photograph. They live in arid areas. In parts of Egypt, they have reputedly fed on the mummified flesh of the pharaohs, being omnivorous in their feeding habits.

Some spiny mice (*Acomys* species) are much lighter in coloration than others. Their head is long and triangular.

A number of African mice have stripes running down the sides of their bodies. This is the African striped grass mouse (*Rhabdomys pumilio*).

rattus) or brown rat (*R. norvegicus*), with between three and six young being born in a litter. The breeding season is often influenced by climate, typically being depressed during periods of dry weather.

A similar division can be made in the case of the house mouse (*Mus musculus*), in which there are both wild and commensal populations. Indeed, it is unlikely that these mice would have spread even across Europe from their Asiatic homeland without inadvertent human assistance.

There is even a case for splitting *Mus musculus* into several species through its present range. Certainly, a clear division is often possible between commensal and free-living individuals, the former usually being darker and having longer tails. The truly wild form also tends to be more nocturnal in its habits. Both build a nest of grasses or of shredded paper and similar items, depending on their environment.

Female mice typically produce a litter of about six pups, which are helpless at birth. Mortality up to the point of weaning, at about 3 weeks old, can be quite high. The young mice may themselves be mature in a further 2 weeks.

Given warm conditions and access to a constant supply of food, commensal mice may breed for much of the year, rather than just during the summer months. This extended breeding period has undoubtedly helped these mice to colonize and establish themselves in areas far outside their native home.

Although mice have no obvious means of defending themselves against predators, apart from their sharp teeth, the modifications to the coat seen in the case of the spiny mice (*Acomys* species) do seem to provide an effective deterrent against attack by owls, which are one of the major predators on mice.

Spiny mice are represented in arid areas of the Middle East, southern Asia and Africa. Their backs and tails are covered with thick, hard spines. Young spiny mice are often born with their eyes open, and they develop quickly. They will be independent by 2 weeks old.

While most members of the Murinae are fairly subdued in terms of coloration, a few species, notably striped grass mice (*Lemniscomys* species), have patterned coats. These mice are restricted mainly to Africa, where they live in grassland, and the stripes running down their bodies provide disruptive camouflage by breaking up their body shape.

The amazing diversity that exists among members of the Murinae is undoubtedly one of the reasons for their success. However, much still remains to be learnt about the habits of many species and the key roles which they play in the ecosystems where they occur.

Blind mole rats • Subfamily Spalacinae

This group of rodents is well adapted for a subterranean existence, even to the extent of being totally blind. Although blind mole rats do have eyes, they are concealed beneath the skin. Their ear flaps are almost absent as well. Blind mole rats use their powerful feet and large incisor teeth to construct their burrows.

This subfamily is represented by three species, which are found in the drier parts of Europe, western Asia and the Middle East. Blind mole rats do not inhabit areas of true desert, however, because their tunnels would collapse in this terrain. They have been known to burrow to over 4 m (4½ yd) below ground but their tunnels usually occur much closer to the surface.

Individuals live apart in their own system of tunnels and chambers, only coming together during the breeding period. At this time, the female digs a breeding mound, at the centre of which is a nesting area lined with grass. Males excavate smaller mounds in the vicinity of a female and these connect with her tunnels.

Young mole rats are naked at birth and then grow a thick coat of grey fur. This is in marked contrast to the slick, short hair seen in adults. The youngsters will be independent soon after they are 1 month old.

Blind mole rats are entirely vegetarian in their feeding habits, seeking roots, tubers and bulbs as they tunnel through the soil. In areas where root crops are being cultivated, they can prove to be major pests. As much as 18 kg (40 lb) of sugar beet and potatoes have been unearthed in a single blind mole rat store. If a blind mole rat comes to the surface at night, where it is very vulnerable to nocturnal predators, it may pause to eat vegetation.

Central Asiatic mole rats • Subfamily Myospalacinae

These mole rats are similar in size to those of the Spalacinae but can be distinguished from them by the short tail. They also have small eyes and more variable coloration, ranging from shades of grey to russet.

The burrowing ability of these rodents, which are sometimes called zokors, is phenomenal. They can dig at a rate equivalent to 1 m (3 ft) every 17 minutes, pausing to shovel earth to the surface after completing a stretch of tunnel up to 5 m (15 ft) long. It is possible to track their progress below ground by observing these mounds. Zokors have both toilet and fodder-storage chambers, as well as a nesting site, within their network of burrows. Breeding appears to take place in the spring and up to six youngsters may be born in a litter.

African pouched rats • Subfamily Cricetomyinae

There are three genera in this group. The African pouched rats (*Saccostomus* species) themselves look rather like hamsters in some respects, having wide, powerful heads, small ears and relatively short tails. As their name suggests, these rats also have cheek pouches, which extend from the lips to the shoulder region. They are solitary by nature and they use their pouches to carry seeds and other food back to their burrows.

The long-tailed pouch rat (*Beamys hindei*) has a longer tail than the African pouched rats and this is flattened along its length, with sharp sides. This rat inhabits areas of woodland.

The largest members of this subfamily are the African giant pouched rats (*Cricetomys* species). These occur over the largest area, with *C.*

gambianus ranging from Senegal to South Africa. Their heads are quite narrow and topped with large, pointed ears, which are almost totally hairless.

Weighing about 1.5 kg (3¼ lb) on average, African giant pouched rats are often hunted as a source of food. They live in separate burrows, which may contain rather strange, inedible items, including coins. Forests and scrubland are the regular haunts, where they become active after dark. Females typically have four youngsters, after a gestation period lasting about 4 weeks.

African swamp rats • Subfamily Otomyinae

These rats, which are found in Africa, south of the Sahara, are not unlike voles, in terms of both appearance and life style. Perhaps their most striking characteristic is their coat, which is long and thick and generally brownish in colour. Both the ears and eyes are quite small and the tail is relatively short compared with the body.

Members of this subfamily are sometimes described as groove-toothed rats because of the presence of a groove running down each of their incisor teeth. They generally live in wet areas where there are tall grasses, using their high-crowned molars to feed on the vegetation. Some nest in underground burrows, while others use runways through the dense grass to provide access to a nest in a tussock. They can swim short distances if necessary and may remain submerged for brief periods if they are threatened by predators.

African swamp rats are less prolific breeders than voles and females produce just two offspring per litter on average. Their young are born in a relatively advanced state of development, with their eyes open.

One species, *Otomys unisulcatus*, has become adapted to a far more arid environment than other African swamp rats. It is restricted to Cape Province in South Africa. It often constructs subterranean tunnels and lines the chambers with vegetation.

Crested rat • Subfamily Lophiomyinae

The sole species in this subfamily, *Lophiomys imhausi*, is confined to northeastern Africa, ranging from Sudan to Kenya. It is also known as the maned rat, because of the unusual and distinctive mane of hair running along its back and part of the tail. The mane itself is dark, with shorter, lighter-coloured fur on either side of it. The tail is bushy and ends in a white tip.

When threatened, the crested rat erects the hair running down its back, which may render it more fearsome to a potential predator. It may create the impression of a porcupine, which possibly helps to prevent it being attacked. A number of other anatomical features are unique to the crested rat. It is, for example, the only rodent where the temporal fossae of the skull are covered by bone.

The arrangement of the toes reveals that crested rats are tree-living rodents, with the innermost digits on each hind foot working in opposition to the other toes, allowing them to grip on to branches.

Vegetarian in its feeding habits, the crested rat squats when eating so that it can use its front paws to hold food. Females are believed to give birth to a single youngster, but little else is known about their breeding habits.

African climbing mice • Subfamily Dendromurinae

These small mice occur across a wide area of Africa. Contrary to their alternative common name of tree mice, however, there is little evidence to show that they live mainly off the ground. In fact, some species are known to live in underground burrows.

African climbing mice are agile climbers, with tails which can be up to a third longer than their bodies and are used for grasping purposes. One species, *Dendromus mystacalis*, may nest well off the ground and it is not unknown for these mice to take over abandoned domed nests of birds for their own use. They have just three digits on each of their front paws.

The breeding period depends to a great extent on where they live. As many as eight young may be born, after a pregnancy of about 25 days. Some species are far more social than others. Nocturnal by nature, African climbing mice eat a wide range of foodstuffs, from seeds to young birds.

Bamboo rats • Subfamily Rhizomyinae

The appearance of these Asiatic rodents varies through their range. For example, their coats are much thicker and softer in the more northerly parts of China, whereas they are thinner and coarser in Malaysia. Bamboo rats live in areas of bamboo, where they burrow among the roots.

Nocturnal in habit, they emerge above ground at night to feed on these grasses. They are vulnerable when out of their burrows and are sometimes caught and eaten by giant pandas (*Ailuropoda melanoleuca*) and other predators. They are unable to run fast and, if cornered, will try to strike first, snapping with their sharp incisors.

Bamboo rats sometimes invade sugar-cane plantations and other areas of agriculture, and eat a variety of vegetables and fruit, as well as bamboo.

The largest species is the Sumatran bamboo rats (*Rhizomys sumatrensis*), which can grow to an overall length of 68 cm (27 in) and may weigh as much as 4 kg (9 lb). They are hunted as a source of food on Sumatra and in South-East Asia, where they are quite widely distributed. Their young develop slowly, and are unlikely to be fully weaned until they are 12 weeks old.

Madagascan rats • Subfamily Nesomyinae

It is unclear whether this group of rodents evolved on Madagascar from a single ancestor whose origins lay in mainland Africa, or whether they represent a series of subsequent colonizations of the island.

The Madagascar giant rat (*Hypogeomys antimena*) is the largest rodent found here, capable of growing to an overall length of 60 cm (24 in). It occurs in western coastal areas around Morondava, living in underground burrows, from which it emerges at night to forage for fruit. The large ears are characteristic of not only this rodent, but also a number of others living on Madagascar, such as the two species of *Macrotarsomys*, which occur on the western side of the island.

Unfortunately, habitat changes, particularly deforestation, are having adverse effects on Madagascar's unique population of rodents. The survival of some species, such as the Madagascar giant rat, could soon be in doubt unless action is taken to conserve them.

Oriental dormice • Subfamily Platacanthomyinae
The two species of Oriental dormice tend to be classified on an anatomical basis, in the family Muridae, rather than with the other dormice, which form the families Gliridae and Selviniidae. The spiny dormouse (*Platacanthomys lasiurus*) shows a clear external similarity to other dormice but its pattern of dentition matches that of other murid rodents, as it has no premolars.

The same distinction applies to the second member of this subfamily, the Chinese pygmy dormouse (*Typhlomys cinereus*). Found in southeast China and northern Vietnam, it looks more like a mouse than a dormouse in terms of appearance, lacking the bushy tip to the tail seen in the spiny dormouse.

Very little is known about the biology of these species. The spiny dormouse is arboreal in its habits and known to damage crops of peppers in some parts of its range.

Australian water rats • Subfamily Hydromyinae
In spite of their name, these rodents occur over a wide area, notably in New Guinea and New Britain to the north of Australia. Beaver rats (*Hydromys* species) are relatively large, growing to a maximum length of 70 cm (28 in) and weighing up to 1.3 kg (3 lb). They are well adapted to an aquatic existence, with a long head, waterproof fur and powerful feet, partially webbed between the toes.

Beaver rats can be quite aggressive predators, feeding on a variety of aquatic creatures, including fish, crustaceans, amphibians, turtles and birds. The seal-like texture of their fur has led to these rats, particularly *H. chrysogaster*, which occurs in Australia, being hunted in some parts of their range. The coat of *H. chrysogaster* may vary in colour, having dark and light areas.

Little is known about some of the other members of this subfamily, such as the one-toothed shrew-mouse (*Mayermys ellermani*), which is apparently restricted to the mountainous region of northeastern New Guinea. This species is the only rodent to have just a pair of molar teeth in its upper and lower jaws. Very slight traces of webbing can be seen between the toes. Shrew-rats (*Rhynchomys* species) from the Philippines also look rather like shrews and feed on invertebrates.

Strangely, not all the rodents grouped as water rats are believed to be primarily aquatic. The mountain water rat (*Parahydromys asper*) lives close to streams in its native New Guinea but appears to feed on invertebrates which it gathers on land, often digging to find its prey. The very wide muzzle and whiskers of this species may help it to find suitable food.

Hamsters • Subfamily Cricetinae

Although many hamsters are short and stocky in appearance, with small tails, the mouse-like hamsters (*Calomyscus* species) belie this popular image. Originating in parts of Asia, this genus is also represented in the Middle East, with the recent discovery in 1991 of a previously unknown species, now described as *C. tsolovi*, in southwest Syria. Large ears and a long tail, which is well furred and terminates in a tuft of hair, are obvious features. These hamsters have no cheek pouches.

A second group, comprising the genus *Cricetulus*, known as rat-like hamsters, extends from southeast Europe to the far east of Asia. They are often found in arid areas, where vegetation is sparse. Rat-like hamsters are short-legged, with tails of medium length. They are so-called because of the shape of their heads, which resemble those of rats. They live underground, taking food back in their large cheek pouches, which can accommodate at least 42 soya beans. They may also eat insects intermittently and, if threatened, are likely to prove highly aggressive, with females often being especially fierce.

The common hamster (*Cricetus cricetus*), which can weigh more than 500 g (17 oz), is the largest species. Its distinctive coloration, with shades of brown on the back and sides, white on the paws and around the muzzle, and jet black underparts accounts for its alternative name of black-bellied hamster. It can grow to about 34 cm (13 in) long, with a short tail that can be 6 cm (2½ in) long. These hamsters used to be common in parts of Europe but their range there has contracted in recent years, although they remain widespread in Asia.

They may burrow to at least 2 m (6½ ft) below the surface and each hamster lives on its own. Within the network of burrows, there will be a store for food. The accumulation of seeds, tubers and other edible items may weigh as much as 90 kg (198 lb)! These hamsters will also eat insects and other invertebrates, such as snails, so they are not entirely without benefit to agricultural areas.

Males venture into the burrows of females at the onset of the breeding season, during spring and summer. Two litters, each consisting of as many as a dozen offspring, are born after a gestation period of about 19 days. They grow very quickly, achieving adult size by the time they are just 8 weeks old.

Although the Syrian hamster (*Mesocricetus auratus*) is sometimes called the golden hamster, it is not actually golden, but rather reddish brown. This appears to be a rare and localized species in the wild, in spite of its widespread familiarity as a pet (see page 32). The glands on the flanks are significant for scent-marking. These hamsters are solitary by nature and occupy their own individual burrows.

The Syrian hamster (*Mesocricetus auratus*) is a solitary rodent, with a very restricted distribution in the wild. It lives in burrows during the day, emerging to forage and feed at night.

There are three other members of this genus. These are more widely distributed through the Middle East and into Asia, with *M. brandti* being present in Turkey. The females in each species have a large number of nipples, varying between 12 and 17. Although they produce about 6 young in a litter, they can have up to 13 on occasions, and suckling this number of offspring is clearly feasible. Members of the *Mesocricetus* species are significantly smaller and have shorter tails than the common hamster (*Cricetus cricetus*).

The final genus in this subfamily is *Phodopus*, whose members are also

The two small dwarf hamsters forming the genus *Phodopus* are found in eastern Asia. The dark black stripe seen here is not apparent in all cases.

At the onset of winter, these dwarf hamsters (*Phodopus* species) may moult their dark coat, which is then replaced by white fur, helping to provide camouflage when snow is likely to be on the ground. This individual is just starting to turn white.

the smallest. These dwarf hamsters, from Mongolia, Siberia and China, may measure no more than 6 cm (2½ in) long, including their short tails. They tend to be greyish with white underparts. They are also called 'small desert hamsters' because they live in arid areas.

Dzungorian hamsters (*P. sungorus*) often have a black stripe which runs down their back. Their coloration alters with the season, becoming white in the winter when snow is likely to be on the ground. The gestation period of these dwarf hamsters is about 18 days, after which an average of five youngsters is born.

Gerbils • Subfamily Gerbillinae
Members of this subfamily are to be found in arid landscapes, ranging from savannas to deserts. The best-known species is undoubtedly the Mongolian gerbil (*Meriones unguiculatus*), which is a popular pet. Other *Meriones* species are better known as jirds rather than gerbils.

In order to escape the intense heat and the risk of dehydration, gerbils live in underground burrows during the day, emerging at night, when the temperature drops and they are less at risk of being detected by predators. The coloration of their coat matches that of their surroundings very closely, providing good camouflage.

Some *Meriones* species are quite social by nature, whereas others are strictly solitary. For much of the year, they subsist on dry seeds and similar food but they will also eat insects, as well as fresh greenstuff. While the Mongolian gerbil is active throughout the year, emerging from its burrow even when the temperature is below freezing, some species rely on food stored in their burrows to sustain themselves when the climate is inhospitable.

The Libyan jird (*M. libycus*) may amass 10 kg (22 lb) of food. These rodents live in family groups, sharing the network of tunnels. As many as nine youngsters may be born to a single female about 30 days after mating. Each female may have four litters during a year and the young will be mature by 4 months of age.

The acute hearing of these rodents enables them to detect danger, such as the wing beats of an approaching owl, allowing them to retreat to cover at an early stage. They have adapted to living in some of the driest parts of the world by conserving water effectively and producing a highly concentrated urine.

The great gerbil (*Rhombomys opimus*) is slightly bigger than the *Meriones* species and occurs in Asia, ranging from Afghanistan to northern China. Social by nature, these gerbils keep only some of their burrow entrances free from snow during the winter and may stay below ground for long periods. There they feed on the large stores of plant matter which they have accumulated, ranging from seeds to roots. They may store as much as 60 kg (132 lb) of food.

These gerbils breed during the spring and summer, with females producing several litters during this period. They are unlikely to live for more than 3 years, particularly in agricultural areas in the former USSR, where they are frequently hunted for damaging crops.

The coloration of Shaw's jird (*Meriones shawi*) matches that of its environment very closely. Living in arid areas, these burrowing rodents may have food storage chambers in their network of tunnels.

The naked-sole gerbils are divided into two genera, based on size, with the larger members comprising the genus *Tatera*. While one species, *Tatera indica* is present in Asia, the remaining species are to be found in Africa, where they are widely distributed. They are so-called because of the lack of fur on the soles of their hind feet. Although they usually forage on all fours, they can bound away on their powerful hind limbs to escape danger.

In the case of the small naked-sole gerbils (*Taterillus* species), some individuals may be as large as species of *Tatera* but they can still be recognized because the soles of their feet have a thin covering of hair. Their breeding period is often closely associated with climate, occurring after the rainy period, when food is most likely to be available.

One of the more unusual members of the Gerbillinae is the fat-tailed gerbil (*Pachyuromys duprasi*), which lives in the northern Sahara desert. It has a short tail, which thickens towards the tip and is darker on its upper surface. This species, which is apparently insectivorous in its feeding habits, is found in areas of the desert where there is vegetation. It has acute hearing and powerful front claws.

The enlargement of the auditory bullae found in desert-dwelling rodents can also be seen in the case of the short-eared rat (*Desmodilliscus brauer*i). It is a small species, found in arid parts of north Africa, and can be distinguished from all other members of the family Muridae by having 10 rather than 12 molar teeth in total; there are only two pairs in its lower jaw.

JERBOAS • FAMILY DIPODIDAE

The development of the hind legs seen in gerbils is apparent to an even greater degree in these desert rodents. Their hind legs generally show a fusion of the tarsal bones to form one thicker, stronger central bone. Also, the number of toes on each hind foot is sometimes reduced to three. Where this reduction does not occur, the first and fifth digits usually play no part in supporting the feet.

In conjunction with the changes in the hind feet, the tail is also elongated, providing balance. It is not unknown for jerboas to leap over 3 m (10 ft) at a single bound to escape predators which they would be unlikely to outpace by running. Under normal circumstances, however, they move at a more sedate pace simply by walking on their hind legs.

In order to prevent their feet sinking into the desert sand, jerboas have tufts of hair present on the soles of their feet, so that their weight can be spread more evenly. These tufts also help to sweep sand out of the way as the jerboa burrows. Hairs are also present over the entrance to the ear canal, serving to prevent sand entering and causing an irritation. In jerboas which use their heads for tunnelling, the nostrils can also be closed off by a fold of skin.

There are approximately 30 different species of jerboa, of which one of the most unusual is the long-eared jerboa (*Euchoreutes naso*). Its ears are about three times as long as its head. This species is found in eastern Asia, in China and Mongolia. Its hind feet do not show the same degree of specialization as in the desert jerboa (*Jaculus jaculus*), with five toes being present on each foot, although the bulk of its weight is supported on the central three digits.

The four-toed jerboa (*allactaga tetradactyla*), found in the coastal zone of Egypt and Libya, has lost the innermost digit on each hind foot and the outermost is greatly reduced. Like other jerboas, this species lives in burrows and is mainly active after dark.

Members of the genus *Allactaga* vary widely in their feeding habits, from being almost entirely vegetarian to highly insectivorous, notably in the case of *A. sibirica*, which feeds mainly on beetles.

The five-toed dwarf jerboa (*Cardiocranius paradoxus*) is a rather aberrant member of the family. The arrangement of the bones in its hind feet is very different, with no fusion of the metatarsals. Also, the first digit of each hind foot is much smaller than the other four. Its scientific name originates from the heart-shaped structure of its skull and, although its ear flaps are small, the internal structure of the tympanic bullae is greatly enlarged. The tail may be a little longer than the body and is narrow at the base, before expanding along its length and then constricting again towards the tip. This species lives in Siberia, Mongolia and parts of China.

Jerboas dig different types of burrow, according to the time of year. The comb-toed jerboa (*Paradipus ctenodactylus*), for example, which is found near the Caspian Sea, uses its forefeet, equipped with powerful claws, to construct its burrows. In summer, when conditions are hot and dry, it digs fairly shallow burrows, perhaps just 1.5 m (5 ft) below the surface, which it occupies for a variable period of time, maybe for as lit-tle as a day. At the onset of winter, a much deeper hibernation chamber is dug, often down to 5 m (16 ft). A carefully protected site is chosen for this purpose and the chamber is lined with grass. Here the jerboa pass-es the winter, usually from the beginning of December through to February, relying on its stores of body fat to sustain itself.

Jerboas can communicate with each other by tapping their feet from within a burrow. They can also call, particularly if in danger. When out foraging for food, comb-toed jerboas may venture as far as 11 km (7 miles), bounding if necessary at speeds equivalent to 11 km/hr (7 mph). Breeding is seasonal, with litters of up to six offspring being produced. Females may have more than one litter in a year but their life span tends to be short. In spite of their remarkable adaptations to this hostile environment, they are still vulnerable both to the climate and predators.

JUMPING MICE AND BIRCH MICE • FAMILY ZAPODIDAE

This family of rodents is found predominantly in the temperate zone of the northern hemisphere. The jumping mice, which form the subfamily Zapodinae, have longer hind legs and tails than the birch mice, which constitute the subfamily Sicistinae. Both subfamilies comprise relatively small mice, varying in weight between 6 and 28 g (¼ and 1 oz).

They are found in a wide range of habitats but tend to occur mainly in areas of low vegetation and woodland, often close to water. Birch mice (*Sicista* species), which are confined to Eurasia, are characterized by their long tails, which can be used for grasping purposes.

Living in cold northern areas, birch mice spend the winter hibernat-ing underground, where they may remain for nearly 7 months. During this period their body weight may fall by half and mortality can be high. Breeding is necessarily compressed into the summer months and only one litter comprising 1–11 youngsters, is generally produced by each female. Pregnancy lasts for 4 weeks and the young are independent after a further 4 weeks. These mice are omnivorous in their feeding habits, eating both plant matter and insects.

Jumping mice tend to move either on all fours or by hopping rather than actually jumping, although they can cover distances of up to 3 m (10 ft) in a single bound if alarmed. These rodents can also swim and may even dive to a depth of 1 m (3 ft) to escape a predator. They eat a wide range of foodstuffs and construct nests from grass in the summer, where the female gives birth. The nests are usually concealed in a tree or similar location, and measure just 10 cm (4 in) in diameter.

In southern parts of their range, woodland jumping mice (*Napaeozapus insignis*) may breed twice but the pups of the second litter will need to put on weight quickly if they are to survive hibernation. Fat stores of 6–10 g (½–⅜ oz) are required for hibernation. During the hibernation the body temperature fall dramatically. In the case of *Zapus* jumping mice, another North American genus, it drops from 37°C (98.5°F) to just 2°C (5.6°F).

The increase in fat occurs rapidly, starting within just 2 weeks of the hibernation period, which usually commences in late September. Only a third of the mice survive to re-emerge in the spring, although they wake up for short periods over the winter, about once every 2 weeks. Once the mice come out of hibernation, towards the end of April, insects form the major part of the diet, to restore body condition.

Not surprisingly, the numbers of these mice in a given area often vary from one year to the next. This is most likely to be a reflection of the availability of food in the autumn, which is the critical period before hibernation begins.

DORMICE • FAMILIES GLIRIDAE AND SELEVINIIDAE

The dormice also have a reputation for hibernating through the winter months in northern areas, and also build up stores of body fat beforehand. They sleep curled up in a ball through this period. With the exception of the unusual mouse-like dormouse (*Myomimus personatus*), dormice have long bushy tails and are not dissimilar to squirrels in appearance. They are well adapted to an arboreal existence and their feet, equipped with small, sharp claws, are able to grip branches securely.

The caecum, where cellulose-digesting bacteria and protozoa are located (see page 60) is not present in the digestive tract of dormice. Even so, they eat a variety of foods of plant origin, including nuts, seeds

Distribution of dormice (Gliridae and Seleviniidae).

Dormice are typically found in woodland areas, feeding and hibernating in the trees.

and fruit. Dormice also consume large quantities of insects, which are a major source of protein, particularly during spring and summer. Hibernation appears to be stimulated by the reduced availability of insects in the autumn.

Soon after waking in spring, dormice will mate and, although some species, like the garden dormouse (*Eliomys quercinus*), only breed once a year, the common dormouse (*Muscardinus avellanarius*) may produce as many as three litters. Calls and whistles may be uttered by both sexes as part of the courtship ritual, helping them to locate each other among the branches.

The female constructs a nest in a bush or tree, where, after a pregnancy lasting between 21 and 28 days, she gives birth. She usually produces 3–9 offspring, which are helpless at first. Their eyes do not open until they are nearly 3 weeks old. They will be independent after a further 3 weeks and may live for over 5 years.

Dormice can be very agile and their acute vision helps them judge distances as they leap from one tree to another. They are recorded as having covered 10 m (11 yd) in a single jump. In areas where fruit is grown, dormice are sometimes considered to be pests, although they occur at lower densities than most other rodents.

Not all dormice hibernate. In some species, it depends on where they live. For example, the forest dormouse (*Dryomys nitedula*) is known not to hibernate in Israel, where it may breed two or three times a year, whereas populations in Europe do hibernate and are likely to have only a single litter.

The forest dormouse (*Dryomys nitedula*) lives in dense forests of central Europe and adjacent parts of Asia. These dormice are agile climbers and rarely descend to the ground.

The garden dormouse (*Eliomys quercinus*) is another dormouse which is commonly seen in forested areas, although it may also live in more open habitat in some parts of its extensive range.

The mouse-like dormouse (*Myomimus personatus*), found in parts of Bulgaria and Turkey, and neighbouring countries, is much more reminiscent of a mouse, in terms of both appearance and life style. Its tail is covered with short hairs. These dormice appear to be terrestrial in their habits and may even live in burrows.

The Japanese dormouse (*Glirurus japonicus*) is confined to the three islands of Honshu, Shikoku and Kyushu. This is the most easterly species, occupying a very isolated position in this part of the world. It hibernates through the winter, sometimes sleeping in human dwellings over this period, or utilizing nesting boxes left for birds.

Dormice are well represented in Africa south of the Sahara, where members of the genus *Graphiurus* are widely distributed. Although they spend much of their time off the ground, they will descend from trees to obtain food, even raiding stores of grain intended for poultry. They eat a varied diet, which may include nuts and seeds as well as insects and small vertebrates.

African dormice have short legs, flattened skulls and whitish underparts, contrasting with the darker fur on their back and sides. The breeding season varies from area to area and females have as many as five offspring in a litter. In some parts of their range, these dormice are believed to hibernate.

The desert dormouse (*Selevinia betpakdalensis*) is the sole member of the family Seleviniidae. It inhabits the desert area around Lake Balkhash in eastern parts of Kazakhstan, where it was discovered in 1939. This dormouse has very soft, dense fur and individual hairs can grow as long as 1 cm (½ in). Short hairs are also present on the tail, which is certainly not bushy. During moulting, which takes about a month to be completed, the old fur is shed first from between the ears and then down the back and sides of the body. Skin appears to be shed at the same time. New hairs regrow very quickly, at the rate of 1 mm (almost ¹⁄₁₆ in) in a day.

The desert dormouse burrows and possibly hibernates for periods underground. It can jump and leap if threatened on the ground and climbs well. Invertebrates appear to figure prominently, if not exclusively, in its diet and it is thought to hunt them mainly at night when temperatures become cooler. The origin of this species is something of a mystery. Nothing has been discovered in terms of fossils to give any clues to its ancestry, although in contrast, the origins of the Gliridae are known to date back nearly 60 million years.

Chapter 8
Caviomorpha: Cavy-like Rodents

The distribution of caviomorphs is centred in South America, where they are present across the continent. None occur in Europe or Australia and their distribution in Asia is confined to the south.

The Caviomorpha is the smallest suborder of rodents, although it is more diverse than the Myomorpha, consisting of 18 families, rather than just five. Caviomorphs vary in size from the gundis (*Ctenodactylus* species), which measure about 17 cm (7 in) long and weigh about 175 g (6 oz), up to the capybara (*Hydrochaeris hydrochaeris*), which is the largest living rodent, approximately 1.3 m (51 in) long and weighing almost 80 kg (176 lb).

The capybara is not the only large rodent in this group. It includes a number of other large species which, in some respects, may fill an evolutionary niche in the New World similar to that occupied by the small antelopes in Africa. They are exemplified by the long-legged mara (*Dolichotis* species) and the agoutis (*Dasyprocta* species), with feet resembling hooves.

Apart from the arrangement of the masseter muscle in the jaw, caviomorphs all typically produce relatively few young, in an advanced state of development, after a much longer gestation period than that found in rodents of other suborders.

NEW WORLD PORCUPINES • FAMILY ERETHIZONTIDAE

Sometimes described as tree porcupines, to distinguish them from their terrestrial Old World relatives, these rodents are very adept climbers, spending much of their time off the ground. Their feet are ideally suited for this purpose, with wide soles and sharp, curved claws to help them maintain their grip.

In the most arboreal species, notably the South American species of the genus *Sphiggurus* and the prehensile-tailed porcupines (*Coendou* species), the first digit on each hind foot is reduced in size and forms part of the foot; this provides better anchorage.

In some species, stabilization of the neck vertebrae is achieved by the fusion of the axis or second vertebra with the adjoining third cervical vertebra, although the precise function of this arrangement is unclear. It may possibly be related in some way to the arrangement of the spines.

The distribution of the spines, which are modified hairs, varies according to species. The thin-spined porcupine (*Chaetomys subspinosus*) has spines along its back that are more reminiscent of bristles, whereas, elsewhere on the body, they are more rigid. The skull of this species is

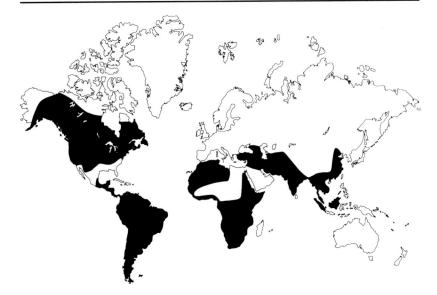

Distribution of cavy-like rodents (Caviomorpha).

reinforced with bone, with the eye sockets being almost completely encircled, unlike the situation in other rodents. The tail is scaly above, with a bare tip and hairs on the underside, but does not act as an extra limb in providing extra support.

The prehensile-tailed porcupines (*Coendou* species), however, do depend on their tails for support. They range from Central America to Argentina. As it climbs a tree, a prehensile-tailed porcupine anchors itself by curling its tail around a branch, releasing it only just before it moves on.

These porcupines often feed on leaves, although they will also eat flowers, fruit and even small vertebrates, such as lizards, on occasions. They do not appear to be prolific breeders and females may only give birth in alternate years. The single youngster is born in the trees and has a covering of hair, rather than spines and soft quills. These become less flexible within days of birth. Prehensile-tailed porcupines produce large offspring that can weigh over 400 g (14 oz) at birth.

The upper Amazonian porcupine (*Echinoprocta rufescens*) apparently has a localized distribution in Colombia. Its protective covering of spines is more marked along its back. The tail, which is quite short, is not used for climbing purposes.

The North American porcupine (*Erethizon dorsatum*) is the most northerly species of porcupine, ranging from northern Mexico to Alaska, often, but not always, being found in wooded areas. It is well protected by a covering of 30,000 or so quills, which can be up to 7.5 cm (3 in) long. These are mixed in among the long guard hairs, while the underside of the body is just covered in hair.

North American porcupines are highly versatile, spending much of their time foraging for food on the ground, but also climbing as high as 18 m (59 ft) on occasions. They can swim well, the air in their quills helping to provide them with buoyancy, and may live in dens, such as caves, burrows and hollow trees, sometimes off the ground.

If trapped, these porcupines will turn and lash out with their spiked tails. They normally lead solitary lives, only coming together during the early winter for breeding. Pregnancy lasts for up to 31 weeks and normally only a single youngster is born. At birth, it is well developed and able to walk, and its eyes are open. Its soft quills are merged with dark hair at this stage but soon harden. Young North American porcupines can survive on their own from just over 2 weeks old but they are probably nursed by their mother for longer.

These porcupines feed mainly on conifers and may strip off the bark to reach the cambium beneath, which results in the death of the tree and makes them unpopular with foresters. They can gnaw food effectively by holding it in their front feet. There are cellulose-digesting bacteria present in the caecum of these porcupines. They may also eat fruit, nuts and other plant matter on occasion, as well as the bones of animals, which provide them with a source of minerals.

CAVIES • FAMILY CAVIDAE

The best-known representative of this family is undoubtedly the domestic guinea pig (*Cavia porcellus*), whose distribution ranges from the northwest of Venezuela to central Chile and which is the original ancestor of today's pet strains.

Distribution of cavies (Caviidae).

There are at least four other members of this genus in the wild, however, all of which are confined to South America. Each has agouti coloration, with alternating light and dark bands running down the individual hairs, giving a greyish or brown effect overall.

Their coat tends to be slightly coarser than that of the domestic varieties but their body shape is distinctive. They have short, bare ears, which are dark in colour, and front teeth equipped with four claws on the toes. There are just three digits on the hind feet.

Cavies tend not to burrow, preferring to hide in clumps of vegetation, and emerge to feed on vegetation at first light and again at dusk. They are quite vocal and associate in groups. Breeding tends to peak during spring. A male will defend a female about to give birth, mating with her soon afterwards. Pregnancy lasts about 9 weeks and the young are miniatures of their parents, able to eat solid food at once and survive on their own, if necessary, after just 5 days.

Cuis (*Galea* species) are very similar in appearance to cavies but are generally paler, with yellowish rather than white teeth. Living at higher

The rock cavy (*Kerodon rupestris*) has a call which sounds like a whistle. Although it spends much of its time on the ground, it may also climb on occasions.

The mara (*Dolichotis*) sometimes rests like a cat, with its front legs tucked up under its body. At other times, these rodents adopt a more vertical stance on their hind legs, wary of danger.

altitudes, they may burrow or take over the abandoned workings of other animals, including other rodents. The females appear to be induced ovulators, so that they come into oestrus only when a male is present. The gestation period lasts just under 8 weeks.

Desert, or mountain, cavies (*Microcavia* species) are also found in montane areas, as their alternative name suggests. Three species are recognized, all of which have a ring of white hair encircling their eyes. They tend to be most active during the day and may climb on occasions. These cavies feed on vegetation as well as fruit and rarely, if ever, drink.

The rock cavy, or moco (*Kerodon rupestris*) is larger than a guinea pig, with a more upright stance. It is confined to eastern Brazil, where it is found in arid areas hiding among the rocks, or sometimes burrowing with its blunt claws. Just one or two offspring are born, after a pregnancy lasting 11 weeks.

Even larger is the Patagonian cavy, or mara (*Dolichotis* species), which is found in southern South America. The larger of the two species, *D. patagonum*, found in central and southern parts of Argentina, is nearly 80 cm (32 in) long, including its short tail, and can weigh 16 kg (35 lb). Maras can run fast in the open countryside and, with their long legs, can reach speeds equivalent to 45 km/h (28 mph).

Habitat changes are spelling a decline in the numbers of maras and other rodents of the pampas area. The introduced European hare (*Lepus capensis*) is more adaptable under such circumstances, placing increasing pressure on the indigenous rodents of the region.

These rodents are very alert and hard to approach. Their claws are hoof-like and assist them when running. Maras shelter in suitable hollows when resting and are normally seen in pairs, although they sometimes form larger groups.

After a gestation period of 11 weeks, two youngsters are usually born. They are in an advanced state of development and will be fully weaned by 11 weeks old.

CAPYBARA • FAMILY HYDROCHAERIDAE

Found in areas of vegetation close to water, the capybara (*Hydrochaeris hydrochaeris*) is quite common in South America east of the Andes. The largest present-day rodent, it relies on water mainly as a retreat from danger, often preferring to feed on vegetation growing on land.

Mating is closely linked with the rainy period and takes place in water. The young are born when there is a flush of fresh vegetation. Pregnancy may last up to 21 weeks, and an average of five, although occasionally as many as eight young, will be born. They weigh 1.5 kg (3 lb) at birth.

Heavy hunting of these rodents has led to their decline in many areas where they were once common. Capybaras are killed for food and in some areas, such as parts of Venezuela, they are now being ranched for

Usually found close to water, the capybara (*Hydrochaeris hydrochaeris*) has open-rooted molar teeth, which grow throughout its life. This helps it to chew up the plant matter on which it feeds. These rodents can sometimes be seen grazing alongside domestic livestock, such as cattle.

Distribution of capybara (Hydrochaeridae).

this purpose, which helps to maintain the natural ecosystem, rather than introducing other species such as cattle, which may not thrive in damp surroundings in any event.

Capybaras are powerful swimmers and can stay submerged for up to 5 minutes if threatened. Males can be recognized by the prominent swollen gland, known as the *morrillo*, on the nose, in front of the eyes. They have a coarse, fairly thin coat and no tail. If danger threatens, a capybara will give a harsh alarm bark, encouraging others nearby to retreat into water. Their unusual name is based on their Indian description, which literally translates as 'master of the grasses'.

COYPU • FAMILY MYOCASTORIDAE

The sole member of this family occurs in southern South America. The coypu (*Myocastor coypus*) is predominantly aquatic and can reach a weight of up to 17 kg (38 lb). It lives in marshy areas, rather than in fast-flowing rivers, where it feeds on aquatic plants. Its hind feet, with webbing between the toes, provide the thrust while it is in the water. On land, the front feet, equipped with powerful claws, are used to dig burrows, which can extend for 15 m (50 ft) or more into the bank.

The demand for coypu fur, or nutria, resulted in these rodents being farmed in a number of countries outside South America and, as a result, wild populations are now established in some regions (see page 22). They are prolific breeders and females often produce two litters of up to 13 young annually, although five tends to be the average.

Distribution of coypu (Myocastoridae).

Coypu that are born early in the year may themselves breed in the summer, when they are only 3 months old, ensuring that a population can grow rapidly. The females' mammary glands are positioned high on the sides of their bodies to enable youngsters to suckle while they are in the water.

HUTIAS • FAMILY CAPROMYIDAE

Confined to the Caribbean islands of Jamaica, Cuba, Puerto Rico, Hispaniola and the Bahamas, the hutias (*Capromys* species) have declined significantly in numbers since the arrival of European settlers, mainly because of habitat clearance and the introduction of predators, such as mongooses and domestic cats. At least half-a-dozen species are believed to have become extinct, some because of hunting by native peoples although one species, *C. nana*, previously known only from its remains, was discovered alive in 1917.

Hutias can climb effectively, although they often remain hidden in suitable hollows underground. They appear not to dig their own burrows. They will eat a variety of foods, including leaves and other plant matter, as well as small vertebrates, and are nocturnal.

Their reproductive rate generally appears to be quite low, with pregnancy lasting over 110 days, after which up to three youngsters may be born. The young are well developed and will themselves start to breed at about 10 months old. Hutias appear to be territorial in the wild and can live for at least 5 years, which probably helps to explain their relatively low rate of reproduction in island surroundings.

Distribution of hutias (Capromyidae).

Hutias (*Capromys* species) have a different digestive system to many rodents, possessing a three-chambered stomach, not unlike that of a ruminant. This enables breakdown of cellulose to occur higher up in the digestive tract, rather than in the caecum.

Some hutias build nests off the ground, located in the relative safety of mangrove trees. These rodents are often arboreal by nature.

PACARANA • FAMILY DINOMYIDAE

Found in the highlands of northern South America, the pacarana (*Dinomys branickii*) has spots and stripes running down its back and flanks. Stockily built, these rodents can weigh as much as 15 kg (33 lb). They have small ears, short legs and a short, thick tail. Their feet are equipped with powerful, sharp claws, which may be used for digging or climbing.

Distribution of pacarana (Dinomyidae).

Pacaranas feed entirely on plant matter. Pregnancy is thought to last between 220 and 280 days, and one or two youngsters are then born. Apparently rare in the wild, the population of these rodents is declining with the destruction of their rainforest home.

PACAS • FAMILY AGOUTIDAE

These shy nocturnal rodents, which can grow to 83 cm (33 in) long and may weigh up to 10 kg (22 lb), are heavily hunted, their flesh being regarded as a delicacy. There are two species of paca (*Cuniculus paca* and *C. taczanowskii*), ranging from Mexico southwards to Paraguay and Peru.

Pacas live in forested areas, burrowing into banks close to water or adopting other suitable retreats, such as caves. They emerge after dark to forage for fruit and plant matter, readily taking to water if danger threatens and sometimes damaging crops in agricultural areas. One or two youngsters are born after a gestation period of about 17 weeks. Pacas have a long life span, at least in captivity, where they have been known to live for over 16 years.

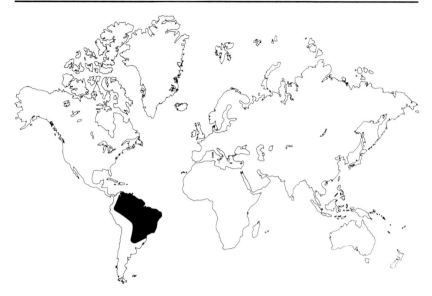

Distribution of pacas (Agoutidae).

AGOUTIS AND ACOUCHIS • FAMILY DASYPROCTIDAE
This is another family which is represented both in Central and South America. The common name 'agouti' should not be confused with the scientific name *Agouti*, which is reserved for the pacas. Despite this, agoutis and acouchis are believed to be closely related to the pacas,

Distribution of agoutis and acouchis (Dasyproctidae).

although they are generally smaller in size, as well as being diurnal.

Their hind feet are considerably larger than their front feet and they can use the strength in their hindquarters to jump up to 2 m (79 in) into the air if surprised on the ground. Neither agoutis nor acouchis climb, relying on concealment to escape detection. Their initial response is to freeze if danger threatens, their dark coloration helping them to merge into the background of their woodland home. If flushed from cover, they can run fast.

Pairs of agoutis appear to maintain individual territories and the male sprays the female with his urine prior to mating. Such behaviour is also seen in other caviomorph species, such as some members of the porcupine family Erethizontidae, and it appears to act as a sexual stimulant.

The female gives birth, typically to twins, after a gestation period of 16 weeks. Although the youngsters are well developed at birth, their mortality can be very high, especially if they become independent at a time when fruit is in short supply.

Resting in this position, with one of its front feet off the ground, is a typical defence posture of an agouti. It may suddenly dart off if it feels directly threatened.

Agoutis (*Dasyprocta* species) sit on their haunches, holding their food in their paws.

The green acouchi (*Myoprocta acouchy*) is so-called because of the greenish tinge to the colour of its upper parts. The other species, the red acouchi, is so-called for a similar reason.

Both agoutis and acouchis have long been hunted for food by the native people in various parts of their range. For this reason, agoutis were introduced to a number of Caribbean islands centuries ago, before the arrival of European settlers. There they evolved into different races, most of which have now become extinct.

CHINCHILLA RATS • FAMILY ABROCOMIDAE

The two species of chinchilla rat (*Abrocoma* species) live in southern parts of South America. They are so-called because they look rather like chinchillas (*Chinchilla* species), but have rat-like tails.

The Chilean chinchilla rat (*A. benetti*) is unique in possessing 17 pairs of ribs, more than any other rodent. Feeding on vegetation, chinchilla rats have a very long digestive tract, to 2.5 m (8 ft) in length, and a correspondingly large caecum. They can grow to 25 cm (10 in) long, with a tail measuring a further 18 cm (7 in).

Their dense, soft fur gives them good protection from the cold in the Andean region, where they can be found at altitudes of more than 3,000 m (9,800 ft). Chinchilla rats live in burrows, which they may share with degus (see page 173), and are nocturnal.

Breeding may take place for much of the year, with a maximum of two youngsters being born after a gestation of about 118 days. Chinchilla rats, also known as *chinchillones*, are trapped for their fur. This may account for the decline in their numbers in some parts of their range, but there are no current data on their populations.

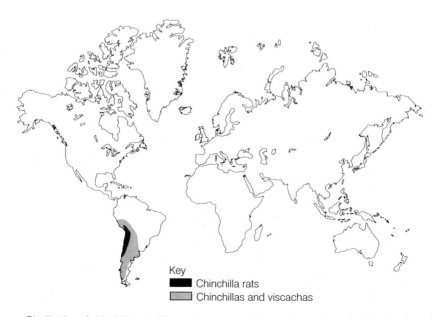

Key
■ Chinchilla rats
▨ Chinchillas and viscachas

Distribution of chinchilla rats (Abrocomidae), chinchillas and viscachas (Chinchillidae).

CHINCHILLAS AND VISCACHAS • FAMILY CHINCHILLIDAE

These Andean rodents have been very heavily hunted in the past for their fur and this has brought chinchillas (*Chinchilla* species) to the verge of extinction. They occur at altitudes of between 3,000 and 4,000 m (9,800 and 13,000 ft), where their soft, thick fur serves to keep them warm.

They feed on vegetation, which they often hold in their front paws, and hide away during the day in suitable crevices. Females can become very aggressive, particularly during the breeding season, which extends from May to November. Two young will be born on average, with pregnancy lasting about 113 days. Wild chinchillas have greyish blue fur, with paler underparts but, as domestication has proceeded, so colour varieties have been developed.

When resting, chinchillas support themselves primarily on their haunches. Their large eyes and big ears serve to warn them of the approach of predators in a landscape where there is little vegetative cover available.

The plains viscacha (*Lagostomus maximus*) now occurs only in Argentina, although it once ranged more widely through South America. This is the largest member of the family, weighing up to 8 kg

The relatively large ears of chinchillas (*Chinchilla* species) are clearly apparent in this domestic variant. Chinchillas have keen hearing. In contrast to the coat itself, the tail of these rodents is covered with coarse hairs.

(18 lb), and has also been heavily hunted. Colonies of up to 50 individuals live in underground networks of burrows, called *viscacheras*. These may remain occupied for centuries and can extend over an area of 600 m² (717 sq. yd).

Males, which are significantly heavier than females, often fight at the start of the breeding season. They frequently have to move some distance away from their viscachera to obtain food, as overgrazing tends to destroy the surrounding vegetation. If threatened, these viscachas can run at speeds equivalent to 40 km/h (25 mph), weaving and turning as they do so, and jumping so that they are harder to catch.

Mountain viscachas (*Lagidium* species) live in rocky, upland areas. They are quite different in appearance from the plains viscacha, possessing long, furry ears. These viscachas do not dig burrows but live in suitable retreats within the rocks, sometimes in quite large colonies. When in danger, they rely on a high-pitched whistle to alert others in the vicinity. Their powerful hind legs ensure that they can climb and jump without difficulty.

A single youngster is born after a gestation period of 140 days. Although able to eat solid food from birth, it will continue suckling for up to 2 months. Life expectancy in the wild is likely to be less than 3 years, particularly where they are hunted for their fur and as a source of food.

DEGUS • FAMILY OCTODONTIDAE

These rodents live in fairly arid areas of South America, ranging to an altitude of over 3,000 m (9,800 ft). They are sometimes known as octodonts, because of the shape of the grinding surfaces of their molar teeth, which resemble the outline of a figure '8'.

Degus (*Octodon* species) have forelimbs longer than hind limbs and well-developed senses. They are about the same size as a rat, with a compact body shape. Active during the day, degus will retreat underground to burrows which they excavate themselves, although they are also capable of climbing if necessary.

When on the move, a degu will hold its tail off the ground. It is able to shed part of the tail, particularly if it is restrained by its tip. It will turn round and actually bite through the tissue and vertebrae, scampering off at once to safety. Blood loss is very minimal and the wound soon heals.

In some parts of their range, degus are considered to be serious pests, causing widespread agricultural damage, particularly as they store food in the autumn for consumption over the winter.

Young degus may sometimes be born with a full coat and their eyes open. About six youngsters form the average litter, with pregnancy lasting approximately 12 weeks. They can survive independently after 2 weeks but usually continue to be suckled for about a month.

The digging abilities of another member of this family, the cururo (*Spalacopus cyanus*), are reinforced by its prominent, protruding incisors. These rodents are blackish in colour, with small ears and eyes,

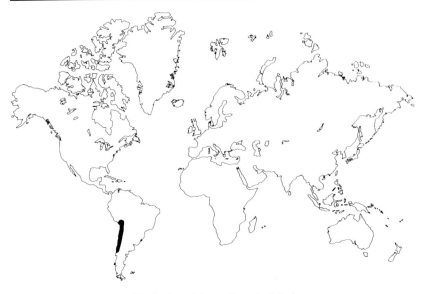

Distribution of degus (Octodontidae).

as befits a burrowing species. Active during the day, the cururo rarely ventures above ground, preferring to eat tubers and parts of the lily, *Leucoryne ixiodes*, which is quite common in the part of central Chile where it is found. When food stores in one area become scarce, cururas expand their network of tunnels. There may be as many as 15 individuals living in colonies.

Little is known about the other members of this family, which include the viscacha rats (*Octomys* species) and the chozchoz (*Octodontomys gliroides*). The latter is known to feed on cactus fruits in the summer, when these are available, and the seeds of acacia pods during the winter. They appear to be less social by nature than degus. Chozchiris sand-bathe, after first urinating on the sand, and males have been observed playing with their young outside the burrow.

TUCO-TUCOS • FAMILY CTENOMYIDAE

This large family, of more than 30 species, comprises the genus *Ctenomys*, whose members closely resemble the North American pocket gophers (Geomyidae) in terms of appearance. They lack external cheek pouches, however, and have a more highly developed covering of hair on their feet. This has given rise to their alternative name of comb rats because these hairs resemble the bristles of a comb. They are used in a similar fashion to remove dirt from the coat.

These rodents are powerfully built for digging purposes, with prominent incisor teeth and sharp claws present on all five toes of each foot. Their tails are short and may have a sensory function, helping them to

Distribution of tuco-tucos (Ctenomyidae).

move backwards through their network of tunnels. These can extend to a depth of more than 60 cm (24 in) below the surface.

Entry points are sealed regularly and are often difficult to locate as a result. By opening and closing entrances, tuco-tucos can keep the temperature in their burrows at an average of 20–22°C (68–70°F), as well as maintaining a fairly constant humidity. If danger threatens, soil is rapidly piled up in the entrance tunnels in order to safeguard the interior of the nest.

Their unusual name derives from the sound of their calls underground. They feed mainly on roots and other vegetation accessible below the surface of the soil and only rarely venture out of their burrows. If they do so, they become very vulnerable to predators, particularly birds of prey.

The breeding season appears to depend on the species, as does the degree of development of the young at birth. Tuco-tucos have been hunted in many agricultural areas for damaging crops, and their numbers have declined accordingly. The extensive networks of tunnels can also be dangerous to horses; they tend to collapse under the animal's weight, causing it to stumble and possibly fracture a limb.

SPINY RATS • FAMILY ECHIMYIDAE

This is the largest caviomorph family, comprising 56 species that range from Central America as far south as Paraguay and southern Brazil. These rodents are quite diverse in terms of appearance, with some having some spiny hairs on their bodies. This is most clearly apparent

Distribution of spiny rats (Echimyidae).

in the case of the thick-spined rat (*Hoplomys gymnurus*), whose spiny coat, best developed over the back, may be as long as 3 cm (1 in). It sheds these spines readily if it is handled roughly. Beneath the spines, there are short, soft hairs.

The most diverse genus within the family are the spiny rats (*Proechimys* species) themselves. Their spines are not as well developed as those of the thick-spined rat. They have long tails, which are easily broken at the fifth caudal vertebra as a means of defence but, once damaged, the tail will not regenerate.

Spiny rats usually live in burrows, emerging at night to forage for plant matter. Although they can climb, they seldom leave the ground, nor do they usually associate closely with people. Their young, born after a gestation period of 4 weeks, are able to move around from birth and are fully independent within 3 weeks.

The arboreal spiny rats (*Echimys* species) live mainly off the ground, scampering around in tree branches rather like squirrels. They also nest in tree hollows, using dry leaves to line the chamber.

Other members of this family have adopted a fossorial life style and tend to spend more time underground. They can be distinguished by their compact body shape and short tail, and include the Brazilian genus *Clyomys*, whose members live in colonies, although much still remains to be learnt about them.

CANE RATS • FAMILY THRYONOMYIDAE

The two species of cane rat, found in Africa, are large, typically weighing up to 7 kg (15 lb), and are somewhat similar in appearance to the

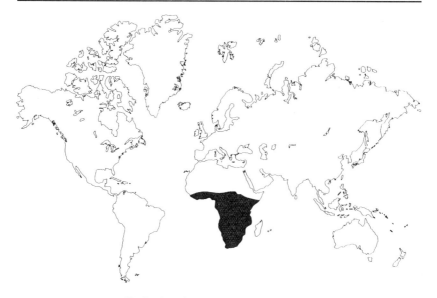

Distribution of cane rats (Thryonomyidae).

coypu (*Myocastor coypus*). The great cane rat (*Thryonomys swinderianus*) is, in fact, semi-aquatic by nature and commonplace south of the Sahara, where it inhabits reedbeds and similar areas of vegetation where it can remain well hidden during the day.

The small cane rat (*T. gregorianus*) is found in wet savanna areas of central Africa and lives in family groups for at least part of the year. It is herbivorous in its feeding habits.

Both these rodents are an important source of food for people in Africa and their meat is in great demand. They are also hunted in agricultural areas for damaging crops, especially sugar cane. When breeding, a female cane rat may have as many as six offspring, which are well developed at birth, after a pregnancy lasting about 14 weeks.

AFRICAN ROCK RAT • FAMILY PETROMYIDAE

The range of the African rock rat, or dassie rat (*Petromus typicus*), is restricted to the southern part of the continent, where it occurs in rocky areas. Its body fur provides good camouflage, matching the colour of the rocks. This makes it difficult for predators to spot them, particularly from the air. When the rock rat retreats under suitable stones for shelter, its flexible rib cage enables it to squeeze through small openings. Its tail also detaches easily for additional protection.

Rock rats may live in pairs, building a nest of vegetation in a secure location. Vegetarian in their feeding habits, they are active during the day. The female gives birth to just one or two offspring in a litter. Her mammary glands are located high on her body so that her young can suckle even when they are in a crevice.

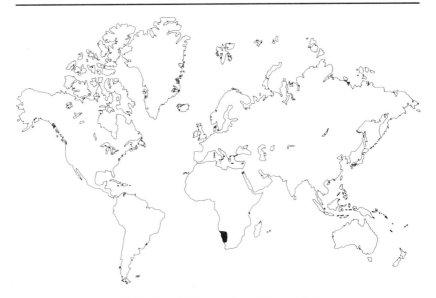

Distribution of African rock rat (Petromyidae).

OLD WORLD PORCUPINES • FAMILY HYSTRICIDAE

Represented in Asia, where they are found on a number of islands, such as Sumatra and Palawan, as well as southern Europe and Africa, these porcupines are adaptable by nature. Unlike their New World relatives, most are mainly terrestrial, seeking protection in burrows rather than climbing.

A range of development in the size and covering of their quills can be seen across the three genera, with the long-tailed porcupine (*Trichys fasciculata*) from Malaysia and adjoining island being considered the most primitive. Its spines are short and cannot be rattled as in other species. The tail, which can measure up to 23 cm (9 in) long, is fragile and easily lost. This is the only Old World species which is predominantly arboreal in its habits.

The brush-tailed porcupines (*Antherurus* species) show greater development of the spines, particularly along the back. The brush at the tip of the tail ends in thick bristles, with the adjoining portion being scaly. These porcupines are nocturnal, emerging from an underground hollow or cave to forage for food after dark. They eat a variety of greenstuff, fruit and invertebrates, as well as carrion, and mark their territories with dung.

Asiatic porcupines of the genus *Hystrix* are well protected, with spines on both their bodies and short tails. The longest quills are seen in the African crested porcupine, *H. cristata*, which can raise its quills to form a crest running along its back. These quills can be 35 cm (14 in) long and there are also specialized rattle quills on the short tail, which can be used to make a hissing noise to deter a potential predator.

The face of porcupines, such as this African porcupine (*Hystrix cristata*), are covered in bristles, with the quills being concentrated on the back and hindquarters. This is why porcupines move backwards towards an attacker.

These large porcupines can weigh up to 35 kg (77 lb). They can swim well and may walk 15 km (9 miles) each night foraging for food, mainly vegetation and other plant matter, although invertebrates are also eaten on occasions. If attacked, they will back directly into their opponent, seeking to drive home their quills, particularly those on the rump, which are short and penetrate the flesh readily. These porcupines do not hibernate, even in European parts of their range, where they have probably been introduced.

Females give birth to about two offspring on average, in a grass-lined chamber in their burrow. Pregnancy lasts about 112 days and the quills of the young porcupines, formed from bristles which start to harden after birth, soon become apparent.

Porcupines are still killed by large cats, especially lions (*Panthera leo*), in spite of their formidable array of quills. They are also hunted by people for meat but, nevertheless, are the longest lived of all rodents, with a life expectancy in the wild which may exceed 15 years.

GUNDIS • FAMILY CTENODACTYLIDAE

The two species of gundi (*Ctenodactylus*) are found in north Africa, ranging from Morocco to Libya. Three other monotypic genera occur in adjoining parts of north Africa. There is still doubt as to where these rodents should be placed taxonomically, because their jaw structure is more akin to the sciuromorphs, whereas the jaw musculature corresponds to the caviomorph pattern.

Some, such as Speke's pectinator (*Pectinator spekei*), which occurs in parts of Somalia and Ethiopia, actually look rather like a squirrel, with its bushy tail. Gundis themselves, however, more closely resemble guinea pigs (*Cavia* species), although they do have tails. Their legs are short, enabling them to slip into gaps between rocks, and the pale coloration of their dense fur helps to conceal their presence. Their rib cage is flexible and their large eyes help them to adapt quickly when venturing above ground into bright sunlight.

They may be forced to travel long distances to find vegetation to eat, spending the early hours of the morning sunbathing on a rock to absorb the heat, before the sun rises. As desert rodents, their kidneys are very effective at re-absorbing water and they do not appear to drink, obtaining moisture from their food. Social by nature, gundis live in colonies and communicate with each other by a series of cheeps and other calls, which can be detected easily by their keen hearing.

Although they are able to move from birth, young remain hidden in crevices while the adults forage, calling their mothers back to them. Pregnancy lasts 8 weeks and up to three young form the litter. Females may give birth several times in quite rapid succession.

It is the presence of combs of bristly hair which account for the scientific name of this family, which is derived from the Greek words *ktenos* and *daktyles*, meaning 'comb-fingered'. These bristles are apparent directly above the claws and the two inner toes of each hind foot.

AFRICAN MOLE RATS • FAMILY BATHYERGIDAE

The amount of hair on African mole rats varies according to the genus, with the Cape mole rat (*Georychus capensis*) having a relatively long coat compared with the short, velvety coat associated with the common mole rat (*Cryptomys* species). This is widespread through much of Africa, south of the Sahara, where soil conditions are suitable for burrowing.

Ears and eyes are scaled down in the case of all five genera in this family, and the limbs are short but powerful. The length of the claws is variable, although the front claws tend to be longer in all cases.

The dune mole rats (*Bathyergus* species), from the southern part of Africa, rely on their forefeet for digging, rather than their incisors, in contrast to other genera, and deposit mounds of excavated soil on the soil surface. It is difficult to distinguish the species in areas where more than one type of mole rat is to be found.

More rats feed underground, eating bulbs and similar parts of plants which they encounter. They rarely show themselves above ground, and the silvery mole rat (*Heliophobius argenteocinereus*) is solitary by nature.

Mole rats give birth to two to four young after a pregnancy of 8 weeks or so. In many parts of their range, they are considered to be crop pests and are hunted accordingly. Their burrowing activities can also be dangerous: trains have been derailed when the weight of the locomotive has caused the rails to collapse into their network of burrows.

Checklist of Rodent Families

While there is no universally accepted agreement on the taxonomy of rodents, this list will give an overview of the families that are generally recognized.

SUBORDER SCIUROMORPHA • SQUIRREL–LIKE RODENTS
Mountain beaver   Family Aplodontidae
Squirrels   Family Sciuridae
Scaly-tailed squirrels   Family Anomaluridae
Springhare   Family Pedetidae
Beavers   Family Castoridae
Pocket gophers   Family Geomyidae
Pocket mice   Family Heteromyidae

SUBORDER MYOMORPHA • MOUSE–LIKE RODENTS
Rats and mice   Family Muridae
Jerboas   Family Dipodidae
Jumping mice and birch mice   Family Zapodidae
Dormice   Family Gliridae; Seleviniidae

SUBORDER CAVIOMORPHA • CAVY–LIKE RODENTS
New World porcupines   Family Erethizontidae
Cavies   Family Caviidae
Capybara   Family Hydrochoeridae
Coypu   Family Myocastoridae
Hutias   Family Capromyidae
Pacarana   Family Dinomyidae
Pacas   Family Agoutidae
Agoutis and acouchis   Family Dasyproctidae
Chinchilla rats   Family Abrocomidae
Chinchillas and viscachas   Family Chinchillidae
Degus   Family Octodontidae
Tuco-tucos   Family Ctenomyidae
Spiny rats   Family Echimyidae
Cane rats   Family Thryonomyidae

African rock rat   Family Petromyidae
Old World porcupines   Family Hystricidae
Gundis   Family Ctenodactylidae
African mole rats   Family Bathyergidae

SUMMARY
Sciuromorpha: 7 families; 65 genera; 377 species
Myomorpha: 5 families; 264 genera; 1,137 species
Caviomorpha: 18 families; 60 genera; 118 species

Further Reading

Although some of the titles in this list are out of print, they should be available from a specialist natural history book dealer or possibly from a public library.

Alderton, D. (1986), *A Petkeeper's Guide to Hamsters and Gerbils*, Salamander Books, London.

Anderson, S. and Jones, J.K. (eds) (1967), *Recent Mammals of the World: A Synopsis of Families*, The Ronald Press Company, New York.

Bright, P. and Morris, P. (1992), *The Dormouse*, The Mammal Society, London.

Clutton-Brock, J. (1987), *A Natural History of Domesticated Animals*, Cambridge University Press, Cambridge.

Corbet, G.B. (1966), *The Terrestrial Mammals of Western Europe*, G.T. Foulis & Co., London.

Day, D. (1981), *The Domesday Book of Animals*, Ebury Press, London.

Delany, M.J. (1975), *The Rodents of Uganda*, Trustees of the British Museum (Natural History), London.

Flowerdew, J. (1984), *Woodmice and Yellow-necked Mice*, Anthony Nelson, Oswestry.

Flowerdew, J. (1993), *Mice and Voles*, Whittet Books, London.

Flowerdew, J.R., Gurnell, J. and Gipps, J.H.W. (eds) (1985), *The Ecology of Woodland Rodents, Bank Voles and Wood Mice*, published for the Zoological Society of London by Clarendon Press, Oxford.

Golley, F.B., Petrusewicz, K. and Ryszkowski, L. (eds) (1975), *Small Mammals: Their Productivity and Population Dynamics*, Cambridge University Press, Cambridge.

Grzimek, H.C. (1984), *Grzimek's Animal Life Encyclopedia*, Vol. 11, Van Nostrand Reinhold, New York.

Grzimek, H.C. (1990), *Grzimek's Encyclopedia: Mammals*, Vol. 3, McGraw-Hill, New York.

Gurnell, J. (1987), *The Natural History of Squirrels*, Christopher Helm, London.

Halstead, L.B. (1978), *The Evolution of Mammals*, Peter Lowe, London.

Haltenorth, T. and Diller, H. (1980), *A Field Guide to the Mammals of Africa including Madagascar*, Collins, London.

Hanney, P.W. (1975), *Rodents: Their Lives and Habits*, David Charles, London.

Hendrickson, R. (1988), *More Cunning than Man: A Social History of Rats and Men*, Dorset Press, New York.

Henwood, C. (1985), *Rodents in Captivity*, Ian Henry Publications, Hornchurch.

Holm, J. (1987), *Squirrels*, Whittet Books, London.

Hufnagl, E. (1972), *Libyan Mammals*, The Oleander Press, Wisconsin.

Keast, A., Erk, F.C. and Glass, B. (eds) (1972), *Evolution, Mammals and Southern Continents*, State University of New York Press, Albany, New York.

Kermack, D.M. and Kermack, K.A. (eds) (1971), *Early Mammals*, published for the Linnean Society of London by Academic Press, London.

Kingdom, J. (1974), *East African Mammals*, Vol. IIB, Academic Press, London.

Kurten, B. (1971), *The Age of Mammals*, Weidenfeld & Nicolson, London.

Laidler, K. (1980), *Squirrels in Britain*, David & Charles, London.

Lever, Sir Christopher (1985), *Naturalized Mammals of the World*, Longman, London.

Lidicker, W.Z. Jr. (ed) (1989), *Rodents: A World Survey of Species of Conservation Concern*, I.U.C.N., Gland.

MacDonald, D. (ed) (1984), *The Encyclopedia of Mammals*, Vol. 2, George Allen & Unwin, London.

McFarland, D. (ed) (1987), *The Oxford Companion to Animal Behaviour*, Oxford University Press, Oxford.

McKay, J. (1991), *The New Hamster Handbook*, Blandford, London.

Martin, P.S. and Klein, R.G. (eds) (1984), *Quaternary Extinctions: A Prehistoric Revolution*, The University of Arizona Press, Tucson, Arizona.

Matthews, L. Harrison (1969), *The Life of Mammals*, Vol. II, Weidenfeld & Nicolson, London.

Novak, R.M. and Paradiso, J.L. (1983), *Walker's Mammals of the World*, Vols. I and II, The Johns Hopkins University Press, Baltimore.

Read, M. and Allsop, J. (1994), *The Barn Owl*, Blandford, London.

Rowlands, I.W. and Weir, B.J. (eds) (1974) *The Biology of Hystricomorph Rodents*, published for The Zoological Society of London by Academic Press, London.

Roze, U. (1989), *The North American Porcupine*, Smithsonian Institution Press, Washington D.C.

Savage, R.J.G. and Long, M.R. (1986), *Mammal Evolution: An Illustrated Guide*, Facts On File, New York.

Scott, W.B. (1962), *A History of Land Mammals in the Western Hemisphere*, Hafner Publishing Company, New York.

Sherman, P.W., Jarvis, J.U.M. and Alexander, R.D. (eds) (1991), *The Biology of the Naked Mole Rat*, Princeton University Press, New Jersey.

Stoddart, D.M. (ed) (1979), *Ecology of Small Mammals*, Chapman & Hall, London.

Wilson, D.E. and Reeder, D.M. (eds) (1993), *Mammal Species of the World: A Taxonomic and Geographic Reference*, Smithsonian Institution Press, Washington D.C.

Index

Page numbers in *italic* refer to illustrations